The Confident Mind

The Confident Mind

A Battle-Tested Guide to Unshakable Performance

Dr. Nate Zinsser

This book is written as a source of information only. While the methods contained in this book have been carefully researched by the author and can and do work, each individual's experience may differ. Nor should the methods contained in this book be considered a substitute for the advice of a qualified professional familiar with a given individual's unique needs and circumstances. The author and the publisher expressly disclaim responsibility for any adverse effects arising from the use or application of the information and methods contained herein. Use of the USMA name does not connote endorsement by the Department of Defense, Department of Army, or the United States Military Academy.

HarperCollins books may be purchased for educational, business, or sales promotional use. For information, please email the Special Markets Department at SPsales@harpercollins.com.

FIRST EDITION

Designed by Jen Overstreet

Library of Congress Cataloging-in-Publication Data

Names: Zinsser, Nathaniel, author.
Title: The confident mind : a battle-tested guide to unshakable performance / Nathaniel Zinsser.
Description: First edition. | New York, NY : Custom House, [2022] | Includes bibliographical references and index.
Identifiers: LCCN 2021030797 (print) | LCCN 2021030798 (ebook) | ISBN 9780063014831 (hardcover) | ISBN 9780063014848 (trade paperback) | ISBN 9780063014855 (ebook)
Subjects: LCSH: Confidence. | Self-confidence. | Performance.
Classification: LCC BF575.S39 Z56 2022 (print) | LCC BF575.S39 (ebook) | DDC 158.1—dc23
LC record available at https://lccn.loc.gov/2021030797
LC ebook record available at https://lccn.loc.gov/2021030798

ISBN 978-0-06-301483-1

22 23 24 25 26 LSC 10 9 8 7 6 5 4 3 2 1

To all who choose to go beyond the everyday normal
and dare to pursue what they might be

Victorious warriors win first and then go to war, while defeated warriors go to war first and then seek to win.

—Sun Tzu, *The Art of War*

Contents

Preface

On August 17, 2011, New York Giant quarterback Eli Manning sat for a live ESPN radio interview after his practice during the Giants training camp. When asked if he was a "Top 10, Top 5" quarterback, Manning said, "I think I am." And then when asked specifically if he was on the same level as New England Patriots quarterback Tom Brady, Manning paused and then said, "Yeah, I consider myself in that class . . . and Tom Brady is a great quarterback."

Manning's statements touched off a torrent of media hysteria. Columnists and bloggers wrote at length about how indefensible Manning's opinion was. How in the world could Manning, with only one Super Bowl championship and MVP award and only two Pro Bowl appearances on his résumé, compare himself to Brady, the six-time Pro Bowler, three-time champion, and two-time NFL MVP? Brady was coming off an excellent 2010 season, throwing thirty-six touchdown passes and only four interceptions, while Manning had thrown a league-high twenty-five interceptions. How could Manning think of himself as Brady's peer?

The answer to that question goes right to the heart of human performance: Eli Manning believes he's as good as any QB in the league *because he knows he has to believe it.* He understands what all champions either know intuitively or have learned during their careers: a performer has no choice but to be totally confident in him- or herself if the true goal is to perform at their top level.

Confidence makes one's peak performance *possible,* and that's why it's of such great importance to anyone who has to step into an arena and deliver their best. Think for a moment about Eli's reality. Most fall and winter Sunday afternoons, he's onstage in front of eighty thousand spectators and millions more watching on TV; his every action on the field (and plenty of them on the sideline) will be analyzed, judged, and criticized by football experts and casual fans alike. If he doesn't have the conviction that he can do his job as well as anyone else (even the guy many consider to be the greatest of all time), then he invites uncertainty, hesitation, tension, and mediocrity into his game. Without that level of confidence, Eli Manning would never play as well as he is capable.

And Manning isn't alone. *Every* quarterback in the NFL has to have that same level of confidence to play his best. In fact, *every* contestant *in any other competitive pursuit* needs it just as much to maximize his or her performance. I'm not just referring to those relatively few individuals who compete in college, professional, or Olympic sports: I'm describing anyone who is striving to achieve success in any field. No matter what "game" you happen to play, you perform best in that state of certainty where you no longer think about how you will hit the ball, throw the ball, or make the move/speech/proposal or about what the implications of winning and losing might be. All those thoughts interfere with (1) your perception of the situation (like the flight of the ball or the movement of an opponent or the understanding of a customer), (2) your automatic recall from your stored experiences of the proper response, and (3) your unconscious instructions to your muscles and joints about how precisely to contract and relax in sequence to make the right move or the right comment at the right instant. Whether your game involves instantly reading a hostile defense and delivering a football to the right spot, returning an opponent's serve, or delivering a sales pitch to a roomful of skeptical prospects, you perform more consistently at the top of your ability when you are so certain about yourself, so confident in yourself, that your stream-of-conscious thoughts slow down to the barest minimum.

Back to Eli and his confident assertion that he was in the same class as Tom Brady. Fast-forward from that training camp interview in August 2011 to February 5, 2012, to the conclusion of that season's Super Bowl. Eli Manning is standing at the center of Lucas Oil Stadium in Indianapolis hoisting the championship trophy and receiving his second Super Bowl MVP award. Manning's New York Giants have just come from behind to defeat Tom Brady's favored New England Patriots. In the closing minutes of the fourth quarter, with the Giants trailing, Manning engineered the 88-yard game-winning drive, making four clutch throws, including a 38-yard pinpoint completion to a tightly covered receiver, unanimously regarded as the "play of the game." Eli Manning showed the world that on that field, on that day, he was indeed a "Top 10, Top 5 quarterback," and that his statement the previous summer was simply the honest expression of a confident competitor.

Now here's a little secret . . . Eli Manning didn't always have that level of total confidence. Despite being the number one pick in the 2003 NFL draft, he had a rough transition from college to the pros, and many questioned whether he'd ever live up to the high expectations that come with being a first-round selection and lead his team to a championship. But in March 2007, Eli Manning started working with me for the express purpose of "becoming a stronger, more confident leader" with a "swagger to match his conscientious preparation." Eleven months later, after diligently working the process of building, protecting, and applying his confidence, Eli Manning led the Giants to a victory in Super Bowl XLII (over Tom Brady's undefeated and heavily favored New England Patriots). All season long he had been turning heads, with sportswriters and sportscasters noting, "This is a different Eli Manning."

So when August 2011 came around and Eli Manning was asked if he was in Tom Brady's class, I wasn't surprised by his answer. By that time Eli had won what we will call the *First Victory*—the conviction that he was good enough to play at a high level on any field against any opponent. He had been exercising his confidence muscles for over four years by then, and despite having had to learn two new offensive systems because of two head coaching changes, despite two recent losing seasons, and despite a revolving door of offensive linemen and other teammates, Eli Manning believed he was as good as any player in his position. He had won the victory in his heart and mind, which gave him the best chance to win on the field in the toughest conditions.

Football experts are still debating whether Eli Manning is indeed a "Top 10, Top 5 quarterback." Arguments about players go on endlessly. What isn't up for debate is that Eli performed at the highest level in a very competitive profession's most demanding and important position for many years until his retirement in 2020. He made the best of his talent and his preparation by building his confidence, protecting that confidence, and playing confidently. He became as good as he could be. The real question is about *you*. Are you as good at your job, your profession, your passion, as you could be? Would your life be different if you won your own First Victory and had Eli's level of confidence (not his arm, not his football IQ, just his confidence)? I'm pretty sure your answer is yes. In the pages ahead, you will find what you're looking for.

INTRODUCTION

What Confidence Is and Isn't

Stoney Portis left his hometown of Niederwald, Texas (population 576), to begin his forty-seven-month West Point "experience" in the summer of 2000. Upon arrival at the banks of the Hudson River he told his cadet team leader that he wanted to continue competing as a powerlifter, because he loved the simple challenge of pushing himself to discover just how much iron he could move. Stoney's team leader sent him straightaway to my office, where, under my supervision and the direct instruction of trainer Dave Czesniuk, Stoney learned, practiced, and mastered mental skills that would enable him to step into any competitive arena and release every ounce of strength and every detail of technique that he built up through his diligent training. By the time he graduated from West Point in 2004 as captain of the West Point Powerlifting team, Stoney Portis benched 345, squatted 465, and deadlifted 505 while weighing only 148 pounds. Five years later, he called upon those same skills to succeed in lethal ground combat in Afghanistan.

Portis's name might be familiar if you saw the intense 2020 movie *The Outpost* or read journalist Jake Tapper's remarkable 2012 book of the same title, upon which the film was based. The "outpost" was Combat Outpost Keating, established by the US Army in 2006 in the Nuristan Province of eastern Afghanistan as part of the US-led coalition strategy to halt the flow of insurgents and weapons over the border from neighboring Pakistan, but unfortunately it was situated deep in a valley surrounded by high mountains where it was vulnerable to enemy fire from multiple positions. Over the next three years it became known, with typical military gallows humor, as "Camp Custer," a place where a massacre could happen at any time. This was the location of then Captain Stoney Portis's command, where Bravo troop, Third Squadron, of the US Army's Sixty-First Cavalry Regiment was stationed on October 3, 2009.

At 0600 local time that morning, Combat Outpost Keating came under attack, but as fate would have it, Captain Portis was thirty kilometers away at Forward Operating Base Bostick, the unit's headquarters, where he had flown two days earlier to coordinate plans to close Camp Keating. Portis got the grim news that his fifty-three soldiers at Keating were taking heavy mortar, rocket-propelled-grenade, and machine-gun fire from the Taliban. By 0830 Captain Portis and the six soldiers who had been with him at Bostick were circling above Camp Keating in a Blackhawk helicopter, preparing to land and join the fight on the ground. This was not Portis's first combat action; in 2006 he had been in firefights north of Baghdad, and he took the same steps now that he had taken then to get control of the naturally occurring flood of negative thoughts that all soldiers experience before battle. "There I was in the helicopter thinking 'This is how I'm going to die,'" he recounted to me. "But I stopped that thought, slowed down my breathing, and repeated one of the affirmations I had been using since the day I took command—*I am the leader; I make the decisions when it counts.* Then I pictured exactly where we would land and exactly what each of us would do once we hit the ground. Before I knew it, I was completely relaxed and in my zone." First Victory achieved.

But as often happens, Stoney Portis's preparation did not meet with immediate opportunity. High above Camp Keating, with his Blackhawk running low on fuel and taking enemy fire, the pilot signaled to Portis that the Taliban attackers had taken control of Camp Keating's only landing zone, so they would have to turn around and fly the thirty kilometers back to Bostick, to both refuel and reorganize. Once again, Portis had to control the fears and worries he felt for his beleaguered soldiers who were desperately fighting for their lives that very moment. Maintaining that control was made all the more challenging once he landed at Bostick and ran to assemble a quick reaction force (QRF) of US and Afghan soldiers that he could lead back to Keating. There his fears were worsened when he passed by the pilot of an Apache attack helicopter that, like the Blackhawk Portis had just flown in, had been badly damaged by enemy fire over Camp Keating. Smoking a cigarette and shaking his head, the pilot told Portis, "I don't know how they'll make it."

Despite the utter seriousness of the situation Captain Portis continued finding his zone for the next nine hours, winning one small First Victory at a time by affirming his conviction, slowing his breathing, and keeping his senses locked in. He helped load the QRF into Blackhawks, flew to the nearest available landing zone atop a nearby mountain, and eventually made his way to

Keating with the QRF on foot via a tortuous five-hour descent covering over two thousand vertical feet of difficult, rocky terrain. Throughout that descent he fought through one ambush after another, calling in artillery and air strikes to turn back the waves of Taliban attackers. By the time he reached Camp Keating as darkness fell at roughly 1800 hours Portis had counted over one hundred enemy dead. The fifty-three soldiers of Bravo troop meanwhile, had fought with extraordinary bravery against an estimated force of three hundred Taliban and prevailed, holding Camp Keating for twelve nightmarish hours. Eight members of Bravo troop died in action that day, and twenty-two more were wounded. Two Medals of Honor, the nation's highest award for valor in combat, eleven Silver Stars, (the third-highest award), and nineteen Purple Hearts (for combat wounded) were later awarded to Portis's soldiers. For his own part, Stoney Portis, who told me, "I'm no hero, I was just in the middle," was awarded a Bronze Star.

Stoney Portis's decision to "find his zone" throughout that day in the worst of circumstances reveals one of the many common misunderstandings about confidence. Most people would certainly not decide to be confident and positive about their future when confronted with such a horrible situation. Most people only allow themselves to feel confident when good things are happening. Their inner state is contingent upon outside events, and thus they are trapped on a roller coaster—flying high when life is a bowl of cherries, and wallowing in the depths the rest of the time. If we are to build, maintain, and apply confidence in the real world of human performance, this common misunderstanding and several others equally ineffective must be put to rest.

Let's face it, our society has a problematic, ambivalent relationship with confidence and confident people. Sure, we all know confidence is important, but we also know that if you come across as more than just cautiously confident, you will most likely be labeled as arrogant or conceited or both. Even quiet and professional expressions of confidence, such as Eli Manning's 2011 assertion described in the preface, generate explosions of questioning and criticism. Confidence, it seems, has a downside—it'll put you in an unfavorable light, either as outspokenly conceited and hence unlikable, as lazy and complacent, or (God forbid) both. As a result of this perceived downside many well-intentioned, dedicated, and motivated people decide not to do the necessary mental work (changing the way they think about themselves) that will build and protect their confidence. Better, they think, to be humble and modest, and that means not developing too high an opinion of themselves. Perhaps they remember all too well that loud and boastful person who beat them at something years ago before they developed enough knowledge or skill to be successful. They'll be damned if they ever let themselves become that loud and boastful so-and-so.

But here's what's important: if you are a naturally quiet individual who grew up believing that it was important not to call attention to oneself, doing the mental work to gain confidence isn't going to change you into a conceited braggart. For every loud and confident individual out there (and the media bombard us with generally negative coverage of loud and confident people—from boxer Cassius Clay in the early '60s to mixed martial arts champions Conor McGregor and Ronda Rousey today)—there are just as many equally confident people who are also naturally quiet and reserved. The truth is that you can be very confident on the inside (which you have to be if you want to perform well), and very polite, respectful, and humble on the outside (which you have to be if you want to have any friends). NFL quarterback Drew Brees, who announced his retirement in March 2021, is one such person. Despite being one of the best players at his position and a former Super Bowl MVP, Brees doesn't say much about himself. He let his play and his other good work, like winning the NFL's Man of the Year Award in 2006 for his charitable work after Hurricane Katrina devastated New Orleans, speak for itself. "I'm a very modest person," Brees told interviewer Steve Kroft on *60 Minutes* in 2010. "But I'm also extremely confident. And if you put me in the situation or in the moment, I'm gonna have some swagger, I'm gonna have some cockiness, and there's nothing I think I can't do." Brees clearly has both the internal, private confidence needed for success and the external, public modesty that puts people at ease.

So remember this: you can be powerfully confident without being considered conceited or arrogant. Go ahead and sound off if that's your natural style. But if you happen to be the quiet, more introverted type, rest assured that following this book's program and learning to win your First Victory won't make you any less polite, respectful, and likable.

With that important point in mind, let's continue making confidence simpler, clearer, and easier to understand. In this introduction I'll establish a simple and functional definition of confidence, one that you can use as a guide in your pursuit of success and growth. With that definition in hand you won't scratch your head or furrow your brow when your boss, coach, trainer, or colleague brings the topic of confidence up (in fact you'll know more about it than he or she probably does). Further, you'll know immediately whether you are fully confident at any given moment for any given task.

Next, I will discuss the five biggest popular misconceptions about confidence, the widely held but misleading ideas about confidence that make it hard for people to build it, keep it, and use it. Once we clear the air on all this, the truth about confidence, the truth about achieving your own First Victory, will emerge. Once that happens you'll know when you have it, and even better, when you know you don't have it, you'll know how to get it.

So let's define *confidence* in a useful, practical way.

Ask a dozen people to state their definition of confidence and you'll get a dozen quick and simple answers. "Believing in yourself" and "Knowing you can do something" are two that I've heard hundreds of times over the years. But these and the others like them that I've heard aren't all that helpful. Just what does it mean to "believe in yourself"? What are the components, the processes, the mechanics behind "believing in yourself"? Unless you care to study philosophy for a long, long time, that definition won't be of much use to you. Neither will be the definitions found in most dictionaries. Here are a couple typical ones: *Merriam-Webster* ("America's most trusted online dictionary") defines confidence as "a feeling or consciousness of one's powers or of reliance on one's circumstances." *Cambridge Dictionary* offers up this one: "a feeling of having little doubt about yourself and your abilities." While neither of these definitions are wrong, neither of them, and none of the others I've come across, are particularly useful to a performer because they all neglect one crucial point about human performance. And that point is this: human beings are hardwired to execute any well-learned skill—be it a tennis backhand, a violin solo, the solving of an algebra problem, or the cross-examination of a witness—unconsciously. No matter how complex the skill may be (and indeed the more complex the skill, the more important this is), the execution of that skill proceeds more smoothly and more effectively when analysis, judgment, and all other forms of conscious, deliberate thought are momentarily suspended. You can have all the "consciousness of your power" you want, but if you're still analyzing your every step, judging your every move, and talking to yourself about how you're doing what you're doing, you'll always compromise your real ability. All those conscious, deliberate thoughts take up a sizable portion of your nervous system's capacity to take in task-relevant information, process it quickly (as in instantly), and send the correct response instructions back out to your hands and your feet (if you need to move), or your throat and tongue (if you need to speak). "When we focus too hard on all the little details of that skill," says Sian Beilock, a psychology professor at the University of Chicago for twelve years and now Barnard College president, "we actually disrupt our performance. If we were shuffling quickly down a flight of stairs and I asked you to think about exactly what both your knees were doing as you were moving, there's a good chance you'd end up in a pile at the bottom of the stairs." Real confidence, then, the kind you'll need to be at your best when the heat's on and the consequences matter, is the absence of all that mental chatter and discursive analytical thought.

So my operating definition of confidence (one that will actually help you perform well), is this: a sense of certainty about your ability, which allows you to bypass conscious thought and execute unconsciously.

Break it down with me:

1. a sense of certainty—that feeling of having complete faith . . .
2. about your ability—that you can do something or that you know something . . .
3. which allows you to bypass conscious thought—so well you don't have to think about it . . .
4. and execute unconsciously—so you perform it automatically and instinctively.

Confidence is that feeling that you can do something (or that you know something) so well you don't have to think about how to do it when you're doing it. That skill or knowledge is in you, it's part of you, and it will come out when needed if you let it.

Allow this definition to sink in by considering the various complicated things you do right now without having to think about them. Tying your shoes is one such activity—ten fingers are engaged in a complex series of delicate movements and adjustments; tension is applied or slackened at progressive intervals; and the proper length of untied lace remains at the end. All this is done without conscious deliberation or analysis. You perform this skill (if you're old enough to be reading this book) with absolute confidence. Consider brushing your teeth—the precise angle of the bristles, the proper amount of pressure per stroke, and the sufficient number of strokes per tooth. All these technical aspects of proper tooth-brushing are executed unconsciously, you do them all without thinking, you do them all with complete confidence. Now consider how useful and helpful this same level of unconscious certainty would be when stepping up to receive a tennis serve against a good opponent, or when playing the most complicated part of the piano recital in front of your teacher and best friends, or when sitting down to negotiate with a hard-bargaining customer. For the cadets and soldiers I teach at West Point, this level of unconscious certainty is an absolute necessity before stepping into hostile territory. Getting to that certainty is what Sun Tzu meant by the phrase "the First Victory."

A brief aside here . . . some readers may wonder if they are indeed "good enough," (that is, skilled enough, or smart enough, or prepared enough), to reach that level of certainty. If you're wondering that, please understand this: success in any field—be it sports, the arts, business, science, and certainly the military—requires both confidence AND competence. A supremely confident individual who lacks the required skills will only be partially successful. The college student who has studied only half the material for the final exam and is utterly certain and comfortable in what she knows will probably not ace the exam; she'll do well on the material she did study (because she's confident about it) but lose points on the rest of the exam. Similarly, the football player who neglected his off-season conditioning program will be at a disadvantage once team practices start, no matter how confident he is.

However, the person who has studied all the material to the point of actually knowing it but still, despite all that study, worries that he's missed something and doubts his preparation, will also never ace the exam, because his constant stream of negative mental chatter will prevent his recall of the facts and the details. In like manner the player who has diligently followed the conditioning program to the letter but still holds on to self-doubt lowers his chances of making the team. It's the person who has done enough preparation, who has developed enough competence, and who then decides to feel totally certain about that level of competence, whatever that level may be, who has the best chance of bringing home that A grade or making the team. So how do you know if you've done enough? Simple: if you can perform the sport skills consistently in practice, or play the tough part of the piano recital alone in your home, or answer all the practice test problems when you're with your study group, then you've probably done enough. But very importantly, no matter how much or how little preparation you have done, no matter how much competence you actually have, your performance when it matters will always depend on whether you feel totally certain in whatever level of competence you have achieved. If you truly want to give yourself the best chance of success, then having that unconscious certainty will always be your best choice.

So how can we ensure that we feel certain about our competence? Where does that all-important sense of certainty come from?

The answers to these important questions require a little digging, and the best place to begin that excavation is with some of the common misconceptions about confidence, the ideas and partial truths that influence popular thinking but aren't truly accurate and definitely not helpful. This exploration will bring us to a useful truth about confidence that will help us build it, protect it, and apply it at the right moments.

Misconception #1: Confidence Is a Fixed, Inherited Trait. You Were Born with a Certain Amount of It and There's Not Much You Can Do Beyond That.

This is an unfortunate but popular misunderstanding. I have met too many people who have given into the belief that their confidence is fixed, so no amount of training or practice or experience will affect it. This is, quite obviously, a self-defeating conviction. If you're convinced that nothing can be done to change your confidence, then you won't bother trying and you'll remain right where you are.

The truth of the matter, however, is quite different and a lot more helpful. The high level of confidence seen in outstanding athletes and other performers is not some genetic accident over which they have no control. Instead, confidence is learned. It is the result of a consistently constructive thinking process that allows performers to do two things: (1) retain and benefit from their successful experiences, and (2) release or restructure their less successful experiences. Believing that confidence (or the lack of it) is an inherited gift gives people an easy and convenient excuse for not putting in the time, energy, and effort to improve their thinking process.

The story of American Olympic bobsledder Jill Bakken is an excellent example of how an individual develops confidence through deliberate effort (while also, by the way, remaining respectful and modest amid success at the world-class level of competition). Standing an unimpressive five foot five inches and weighing maybe 130 pounds, with a shy smile and quiet demeanor, she may not immediately come across as a superconfident performer. Having been completely overshadowed during the 2001 Bobsled World Cup season and right up to the 2002 Olympic Trials by the other American bobsled driver, Jean Racine, Jill had very few logical reasons to feel confident about the upcoming 2002 Olympic Games. Racine and her partner had won the 2001 world championship, were the gold medal favorites coming into the Olympics, and had received all kinds of endorsement money (Visa put them in a national TV ad). Everybody had pretty much forgotten about Jill Bakken. But when it came time for the favored Racine team to put it on the line in the 2002 Winter Olympic Games, they folded under the pressure and were out of the medal hunt after their first run. Enter Jill Bakken.

Coming out of nowhere, never considered as a medal contender, Jill drove her sled down the Utah Olympic Park bobsled track and won gold in the first women's Olympic bobsled competition. When she and her partner, Vonetta Flowers, came through the finish line, with the American flags waving and the hometown crowd going crazy with joy, she jumped out of the sled, hugged Vonetta and her coaches, and was ushered over to be interviewed by CBS TV. The very first question that sportscaster Mary Carillo asked Jill Bakken was, "You were the other team, you weren't supposed to be here. How did you do it?" Through the tears of joy Jill looked at Carillo and said rather plainly, "We just had confidence and that's what we had to go with."

As simple as that sentence was, the words had special meaning to me because of what had happened fourteen months earlier. Back in December 2000, amid all the stress and uncertainty of the upcoming World Cup circuit, Jill Bakken and I met in the lobby of a hotel right outside the newly constructed Olympic Park in Park City, Utah, for our first working session in performance psychology. We sat down in a quiet corner of the lobby and I asked her, "Okay, Jill, you've heard me explain what I do and how I help athletes. What would you like to talk about?" Jill looked me in the eye and without the slightest hesitation said, "I could use a lot more confidence." So we spent the next three hours in that hotel lobby talking about what confidence was, what it wasn't, separating all the BS about it from the truth, and outlining some concrete things that Jill could do to build her confidence day by day.

Over the next fourteen months Jill did those things, despite injuries, distractions, and precious little competitive success. We met a few more times over those months, and Jill stayed at it, doing her best to control her thoughts and emotions enough to always maintain the belief that she could win it all. As she worked on the quality of her thoughts and attitude she changed from the athlete who'd said, "I could use a lot more confidence" to the Olympic gold medalist who explained her team's victory with the phrase "We just had confidence."

The moral of the story is simple and encouraging: confidence is a quality that you can develop the same way you develop any other skill, ability, or competency—through practice. Jill Bakken did just that, and that's certainly one of the reasons she's an Olympic champion. So I hope you realize that it really doesn't matter how much confidence you have or don't have right now—you can always build more, just like Jill Bakken did.

Misconception #2: Confidence Is All-Encompassing, so You're Either Confident Across the Board in All Aspects of Life or You're Not Confident at All.

Quite the contrary—confidence is VERY situation specific. You can feel very confident on the basketball court but feel completely insecure in the history classroom and vice versa. Even on the basketball court you can have entirely different levels of confidence for different aspects of the game—shooting free throws versus shooting off the dribble, posting up versus rebounding, et cetera. In the classroom it's no different; almost every high school and college student I've ever met (and plenty of medical and law students) has a subject or two they feel comfortable with and another subject or two they feel rather insecure about.

The moral here is just as simple and just as empowering as the one above: you can develop confidence in any specific aspect of your life that you care to. Hesitant about your ability to deliver a quality formal presentation but totally at ease doing your research? Comfortable with your tennis serve but anxious when it comes to volleys at the net? Confidence in any of these specific areas can be learned and developed.

Misconception #3: Once You Become Confident You'll Stay That Way Forever.

How I wish that this were true, and every one of my students and advisees wishes the same thing. If only confidence was a onetime, "Now I've got it forever" achievement. Unfortunately, quite the opposite is true—confidence is very fragile—and that's why maintaining it requires consistent attention and effort. One of my West Point cadet advisees (Connor Hanafee, Class of 2013) perhaps put it best when he reflected back on his four years of collegiate wrestling: "Fighting self-doubt and building confidence is a perpetual war of attrition, not a decisive, destructive victory." That statement expresses an essential, but perhaps inconvenient, truth. Just as sustained progress in any sport requires the refinement of physical and technical skills, and just as sustained progress in any profession requires continuous learning, so too does the development and maintenance of confidence require consistent attention and effort. Cadet Hanafee put it in appropriate military terms when he contrasted the "decisive, destructive victory," such as the one achieved by the bombings of Japan that ended World War II once and for all, with the "perpetual war of attrition" that continues to this day in Afghanistan despite twenty years of military engagement. Different kind of war, different kind of long-term involvement, different kind of continuous effort to contain and neutralize an enemy that hides in the shadows and attacks relentlessly.

Bob Rotella, the sport psychology expert best known for his work mentoring PGA and LPGA champions, made the same point about the need to continually work on confidence by comparing it to the work seaside communities do to maintain the sand dunes that protect streets and buildings from the sea. The ocean waves continually pound the shore and wear away the dunes slowly but surely. Sometimes the waves are small, so the impact on the dunes is small and only requires minimal maintenance work by local work crews. Other times big storms cause more damage, and more maintenance work is needed. But at no times can the dunes simply be left alone with the idea that once built they will be fine forever. Just like the waves that constantly grind down the shoreline, the pursuit of success in sport and business will present setbacks and obstacles that can beat down the most optimistic and "positive" competitors. Those who succeed are those who are willing to patiently and persistently build and maintain their confidence.

The moral here is again simple and empowering. The simple part is this: the process of achieving one's First Victory never ends. And the empowering part? Most people think all they have to do is achieve it once and then they can stop. They will subsequently get hit in the face with a setback, a big storm that erodes their personal "sand dunes," and they will give up. That means you, the person who understands that you're fighting Connor Hanafee's "perpetual war of attrition" and continues to build confidence over the long term, will have a significant advantage over nearly everyone else and have fewer and fewer real competitors. Advantage you!

Misconception #4: Once You've Achieved Some Success and Once You've Gotten Some Positive Feedback Your Confidence Is Guaranteed to Grow.

Not really . . . the key term in that misconception is "guaranteed." The old saying "nothing succeeds like success" doesn't tell the whole story. Successful high school athletes do not always make an easy transition to college play, despite their years of previous success; successful high school scholars don't always make a successful transition to college academics despite their great high school grades and top SAT scores. Some successful athletes actually LOSE their confidence because their previous successes become a form of pressure from which they cannot escape. So while experiencing success and receiving positive feedback can indeed both work great as sources of confidence, they do so IF and ONLY IF you allow them to (much more on this to come); they do not, however, guarantee it. Why not a guarantee? Because of the way so many athletes and so many performers who experience great success have developed the habit of focusing on their weaknesses completely and remembering only their failures. So it really doesn't matter how much "success" you've had if you don't let it work for you.

My case study of how success does NOT translate into confidence is the experience of well-known TV personality and former NFL defensive end Michael Strahan. Here are a few facts from Strahan's bio: second-round draft pick in 1993, starter at defensive end beginning with his second season in 1995, All-Pro 1997 season where he recorded a league-leading fourteen sacks, and owner of a multimillion-dollar contract. By nearly any analysis, Strahan was a most successful individual and hence had plenty to fuel a high level of confidence. However, in a 2001 *Sports Illustrated* article, as Strahan's New York Giants were heading into Super Bowl XXXV, he told a much different story:

"The thing that haunts all players is self-doubt . . . Toward the end of 1998 I had 10 sacks in 10 games, but I thought I sucked . . . It was like we had no hope."

How is it possible that Michael Strahan, given that he came off a stellar 1997 season and had already accumulated one sack per game in 1998, was not confident ("I thought I sucked" . . . "we had no hope")? The answer lies in how Strahan was thinking about himself during that statistically stellar season. In the same article he described his vision of what it was like to be on the field this way: "I picture chasing the quarterback, almost getting there, not getting there; and then everything goes black." Despite a level of success that any other player would love to have, Strahan's dominant thoughts about his play were about failure ("not getting there"), and they overshadowed, blocked out, and functionally eliminated the great confidence-building power that the memory of all his great plays could have given him. The good news for Michael Strahan (and his team) was that he changed his destructive mental habit by learning to recall and enjoy all the plays in which did succeed, and he went on to have a Hall of Fame career.

The moral of this story is that success in and of itself is not a confidence booster. It's what you do with your success-related thoughts and memories that determines whether you feel confident. You can discount them as unimportant or ignore them altogether as Strahan once did, or you can use them constructively as Strahan learned to do and in the process set yourself up for greater success. Choose wisely!

Misconception #5: Mistakes, Failures, and Negative Feedback Inevitably Destroy, Erode, or Weaken Your Confidence.

If you've been reading carefully up to this point you probably know where I'm going with this one. Sure, mistakes, failures, setbacks, and so on, can indeed give one pause and get one worrying about what might happen next. But just as we saw in the last section how success only builds confidence if you let it, a mistake, even a serious one, only erodes confidence if you let it. It's just as possible to selectively reinterpret that mistake as a learning opportunity, to view that failure as a momentary, isolated incident, and to take any negative comment directed at you as a stimulating challenge. Put bluntly, it doesn't matter how much "failure" you've experienced if you decide to respond to it constructively. And sometimes maybe "responding to it constructively" means you ignore it altogether.

Put yourself in this scenario: You've made it to the Olympics in your selected "sport." You're about to "perform" on the biggest stage of your career. Maybe you're performing your first organ transplant as the surgeon in charge; maybe you're interviewing for the dream job or taking command of the dream project with the crackerjack work team you've always wanted. What you have dreamed of and worked toward for years and years is about to come to pass! You're going through the final preparations, actually "warming up" just minutes before going into that operating room, or that conference room, or onto that "field of friendly strife" and something goes terribly wrong. For some unknown reason you can't get your body loose, or you start drawing a blank about some phase of the operation, or your presentation notes can't be found. Setback with a capital *S*. How would you feel at that moment? How would you feel as you started that performance? Certain enough to perform "unconsciously" or bombarded with worrisome mental chatter?

This was the situation figure skater Ilia Kulik faced for his short program at the 1998 Winter Olympics. His warm-up routine did not go as planned—a few slips, some wobbles, no impressive jumps. But when it was his turn to take the ice and perform in front of the Olympic judges and a worldwide television audience he was near perfect, placing first in the short program en route to winning the gold. Moments after that performance, while still catching his breath, he was ushered to a TV interviewer for one of those up-close-and-personal moments. The dialogue went like this:

SPORTSCASTER: Ilia, the biggest challenge of the short program is landing that first combination and dealing with all that pressure. Were you nervous coming into this program?

KULIK: Yes, the short program is the most nervous part, because there are eight elements and you have to do them all clean or you just lost.

SPORTSCASTER: Tell us about the combination you needed so badly. Tell us how you felt going into it [as a videotape of his flawless first triple jump plays on the monitor behind them].

KULIK: It was quite complicated in the warm-up . . . but I knew I will do this in the program, I knew it 100%. If in your mind you're 100% confident in what you're doing in the program, there's nothing to do in the warm-up.

SPORTSCASTER: [throwing up her hands in a gesture of disbelief]: Where did that confidence come from?

KULIK: [shrugging his shoulders]: I don't know, just from my mind.

The setback that would have rattled most individuals (a poor warm-up prior to the biggest performance of one's life) was no big deal for Ilia Kulik. Instead of lingering on the possible effects of a bad warm-up (easy enough to do) he just thought about nailing each jump of his program ("100% confident in what you're doing in the program"). In his mind the warm-up, his most recent experience and thus the one most likely to influence his attitude, was meaningless. The potentially significant setback, the potential blow to his confidence, ended up being a nonevent. If anything, it only strengthened his determination to succeed in the actual program. Unlike Michael Strahan, whose mind was dominated by thoughts of failure despite plenty of available "success," Ilia Kulik's mind was dominated by thoughts of success despite a recent "failure."

The moral of this story is that "failures," even ones that come at very inopportune times, are not necessarily confidence destroyers. They only do so when you linger on them, review them, replay them. They can indeed bring on worry, doubt, fear, and a host of other negative feelings, or they can serve as cues to us to go back to a personal store of remembered successes. Confidence comes, the Olympic champion tells us, "from my mind" and not from anything that happens outside of that mind.

And Now We Get to the Truth . . .

The experiences presented above, all from real performers striving for success in their chosen fields, bring us to the simple and practical truth about confidence, about winning Sun Tzu's First Victory. That truth is this: confidence has relatively little to do with what actually happens to you, and pretty much everything to do with how you think about what happens to you. Once Jill Bakken ignored the fact that her previous training hadn't included anything about confidence (what had happened) and decided to be a lot more careful and selective about her memories (what she thought about), she became a confident Olympian and eventual champion. Once Michael Strahan replaced his visions of futilely chasing the quarterback with visions of dominating play, his previous self-doubt no longer haunted him. And as Ilia Kulik refused to fixate on an untimely setback he maintained the confidence to skate a gold medal performance.

It's no stretch then, to think of your confidence, that sense of certainty you have about yourself and your abilities, as the sum total of all your thoughts about yourself and your abilities. In the world of human performance, your confidence regarding your sport, or game, or profession is the sum total of all your thoughts about that sport, or game, or profession. Taken further, your confidence regarding any single aspect of your "game" is the total of all the thoughts you have about that aspect (forehand, backhand, first serve, second serve, volleys, etc., for tennis; passing, shooting, checking, etc., for hockey; budgeting, forecasting, employee management, etc., for business). But this total of thoughts isn't a static, once-and-for-all tally. Instead, it changes constantly as each and every new thought and new memory is added to it, making it a "running total," a momentary sum of everything that you've thought about yourself and your abilities, a sum that is always changing depending on (1) how you are thinking at any time, (2) which aspects of your experience you are choosing to focus and linger on at any moment, and (3) how much emotion you invest in which particular thoughts and which particular memories. In that way, human confidence is very much a psychological "bank account," a repository of your thoughts about yourself and what is happening in your life. Just as the balance of any bank account at the end of the day depends upon how much is either deposited into it or withdrawn from it, the psychological bank account of confidence also rises and falls depending on how you are thinking at any moment. "Deposit" into that bank account memories of past successes, memories of progress or improvement, and thoughts about future improvements and accomplishments, and the "balance" grows. "Withdraw" from that account by replaying past setbacks and difficulties, or by fixating on possible future setbacks and difficulties, and the "balance" shrinks. Gaining confidence, protecting confidence, and performing with confidence—winning that First Victory—is all about managing your psychological bank account.

Pause for a moment and consider what's in your own bank account ("What's in your wallet?" goes the TV ad). Honestly ask yourself: When you think about your involvement in your sport or your profession (or in anything else that matters to you), which type of thoughts dominate, memories of mistakes or memories of spot-on execution? Visions of ongoing troubles (like the "early" Michael Strahan), or visions of desired success (which Strahan learned to deliberately "deposit" into his personal bank account)? What exactly are you putting into your bank account? Is your balance growing every day no matter what actually happens in your life or does it fluctuate wildly depending on your most recent performance, test, or evaluation?

Once you understand that every thought and every memory you have about your sport or your profession is affecting your ongoing sense of certainty, you can decide to either take command of how you think or give this command over to the ups and downs, the highs and lows of life. Taking this command over the input into your personal mental bank account will create for yourself an advantage over all those who don't. The ability to do this—to selectively interpret your personal experience so that you mentally retain and benefit from experiences of success, progress, and effort, while simultaneously mentally releasing or restructuring experiences of setbacks and difficulties—is in all of us. It is the primary mental skill upon which the First Victory is won.

Psychologist Viktor Frankl, in his memoir of surviving the Nazi concentration camps of World War II, called this process of selectively interpreting one's personal experience "the last of the human freedoms: to choose one's attitude in any given set of circumstances." Frankl recognized that confidence in the face of very real and life-threatening challenges was an ongoing process, that one's attitude was indeed a constantly changing running total of everything one thought. "Every day, every hour," he observed, "offered the opportunity to make a decision, a decision which determined whether you would or would not submit to those powers which threatened to rob you of your very self, your inner freedom."

The chapters that follow are an exploration of, and a guide to exercising, this "last human freedom." Few of us will ever (thankfully) experience anything remotely comparable to the horrors that Frankl endured during his captivity. His experience, however, is a powerful reminder and vivid testimony to the power we all have—to separate our thoughts and attitude from what happens to us and around us. Each of us can build that sense of certainty about our abilities, and when we do so, we create the platform (the bank account) from which those abilities can be fully expressed. As Max Talbot, a veteran NHL and international hockey player, put it in one of our meetings, "If I do this, I can become really rich!" And he wasn't talking about money.

Chapter One will explain how to set up and start making daily deposits in your own bank account. You can become "rich" too.

CHAPTER ONE

Accepting What You Cannot Change

It was a relatively normal day at the pharmaceutical firm where Ginny Stevens worked as a midlevel executive. She knew there'd be a product presentation to a conference room full of corporate vice presidents today, but she was one of several attendees with no formal role in the actual presentation, so she was her usual calm and relaxed self. All that changed as she left her office and headed to the meeting. Her boss caught up to her just a few strides away from the conference room door and told her that she would be giving the presentation herself. What followed was a moment of utter panic.

Ginny described it to me this way: "I stopped in midstride and my head did such a fast 180 to look back at my boss that I almost hurt my neck. I take pride in being a good employee, so I really couldn't say no. But inside my mind was screaming, 'What!?! Are you kidding me? You want me to do a presentation to a room full of VPs with no forewarning and no preparation?!!' We kept walking toward the conference room and the doors seemed to grow before my eyes, getting bigger and bigger until it seemed like I was entering some huge cathedral where a council of elders was waiting to pass judgment on me. Even though I knew the product pretty well, I had no idea how I was going to pull this off. I just about lost it when the doors opened and I could see those VPs sitting inside waiting for me."

Luckily for Ginny, the VPs took it easy on her and the presentation turned into a pleasant, collaborative conversation. But it might not have gone that way, and Ginny's moment of panic was so upsetting that she knew she had to do something about it. "I hear that you help athletes with their confidence," she said to me, "but I sure need it, too. And so does my entire work team. Can you help me? I don't ever want to feel that way again."

Ginny's story is one I have heard a thousand times: a sudden change of circumstance or situation sends someone into a tailspin of self-doubt. The heart starts pounding, the thoughts start racing, and even everyday perceptions of time, space, and surroundings shift uncomfortably. One moment you are walking to a run-of-the-mill event, and the next moment you are staggering toward what seems like your own execution.

But it doesn't have to be that way. You can armor yourself against any unexpected turn of events and also against all the known and expected difficulties of life by building for yourself a personal mental fortress, a guaranteed foolproof bank that holds your personal confidence account.

How do you build that bank? You start with some strong foundations, and naturally, because this is a mental bank account, your foundations are mental too. There are four mental pillars that we begin with, four factors that affect all human performance. Once you accept them, you'll see the pursuit of excellence a lot more clearly, and you'll have the peace of mind that will help you build a lasting structure.

Maybe you've seen the well-known "Serenity Prayer."

God, grant me the serenity to accept the things I cannot change,
Courage to change the things I can,
And wisdom to know the difference.

I like that word *serenity*. It suggests a certain level of inner peace, a certain level of mental calmness, a stable foundation from which growth and development can happen. For the purpose of confidence, for winning that First Victory, that serenity is the foundation of the fortress that houses your mental bank account. And you establish that foundation of serenity by accepting four realities of human performance that you cannot change. These four pillars are (1) the mind-body connection, (2) human imperfection, (3) the action of the autonomic nervous system, and (4) the delayed returns of continued practice. We can choose to ignore and resist these concrete realities of our human existence, or we can choose to acknowledge them and work with them. The former path leads to stagnation and mediocrity, and the latter path to growth and success. The choice is yours. Let's explore the four pillars of a kick-ass confident attitude.

Pillar #1: The Mind-Body Connection Is Real. Use It or Be Used by It.

The term *mind-body connection* entered the public vocabulary during the late 1960s and early 1970s, a period of considerable social and ideological change. In earlier years, it was generally understood in Western rational scientific thought that the mind and the body were entirely separate and distinct; your thoughts and emotions were the domains of priests and poets, while your body, obeying the physical laws of chemistry and physics, was the domain of mechanistically trained medical doctors. So it really didn't matter what or how you thought—your body, the instrument through which you played your sport or performed at your job, didn't care. That started to change when yoga, meditation, and other Eastern practices began attracting attention and gaining adherents among both the public and the scientific community. Laboratories at traditional bastions of scientific respectability such as Harvard and Stanford undertook significant research efforts to determine whether mental techniques like Transcendental Meditation could affect bodily processes like blood pressure, oxygen consumption, and heart rate. The results were conclusive beyond any doubt—when subjects changed how they thought and cultivated a peaceful, serene emotional state, their bodies did indeed respond with dramatic reductions in these processes. Herbert Benson's classic 1975 bestseller, *The Relaxation Response*, describes these remarkable findings and their implications for health and healing. Conversely, other studies showed that when subjects focused on memories of arguments and other stressful moments, these same bodily processes accelerated (see *Anger Kills*, by Redford Williams).

Despite Benson's work and the hundreds of scientific papers that followed, the idea that one's state of mind has substantial effects on one's physical state and thus on one's performance still hasn't caught on fully. If it had, a lot more people seeking performance success and the resulting satisfaction would be paying a lot more attention to their minute-by-minute thinking habits. Almost fifty years have passed since Benson's first publications, but I bet you'd be astonished by what a simple heart rate monitor reveals were I to connect you and then ask you to remember various experiences. "Recall the feeling of being in a comfortable hot tub" would make your heart rate's line graph plummet and a comfortable sigh escape your lips. "Recall the feeling of being around that teacher in high school who was always criticizing you" would cause the same line to skyrocket. You may be like all the professional athletes and the millions of weekend warriors who carefully monitor their food intake and religiously follow their workout schedules, but who largely ignore the reality that their thinking habits play a significant role in what actually happens—not only when they step onto the court or into the workplace but during every waking minute.

The First Victory begins with accepting and utilizing the connection that decades of mind/body research has established: your conscious thoughts have a huge influence on your performance by the way they shape your mood and in turn affect your physical state.

And it doesn't stop there. Each performance becomes the subject of further conscious thoughts and initiates another round of the cycle. The result is a continuously operating, twenty-four/seven, 365-days-a-year process where your thoughts influence your feelings, which in turn influence your physical state, which in turn influences execution, which is then "thought about." This cycle influences everything we do as human beings from the physically delicate tasks, like writing an essay on a history exam, to the physically demanding tasks, like a full-contact boxing match. We are embodied beings, and this fact of our existence

There's a Connection...

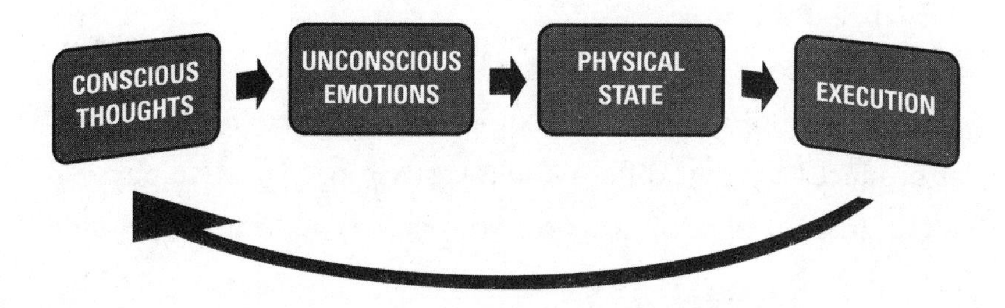

makes the management of our physical state through control of our thoughts and subsequent feelings a prerequisite for optimal performance.

This connection is constantly working to either enhance or degrade your performance; there really is no neutral or middle ground. If your emotional state, driven by a flood of worrisome thoughts, has produced an accelerated heart rate, higher blood pressure, increased muscle tension, tunnel vision, and a cascade of stress hormones, your execution is likely to be compromised,

The Sewer Cycle...

Oh #!#!...
This sucks...
I'm in trouble now...
Don't screw up!

Disappointment
Frustration
Impatience
Worry

High muscle tension
Constricted blood flow
Tunnel vision
"Stress" chemicals

Average to poor

CONSCIOUS THOUGHTS → UNCONSCIOUS EMOTIONS → PHYSICAL STATE → EXECUTION no matter what kind of task you are attempting. I refer to this as the "sewer cycle" (you know what goes down the sewer). You can visualize it this way:

Conversely, if your emotional state is driven by a flood of constructive thoughts (notice I did not say "positive" thoughts), it produces a rather different and vastly more effective physical state. Instead of feeling tense, you feel energized; your vision

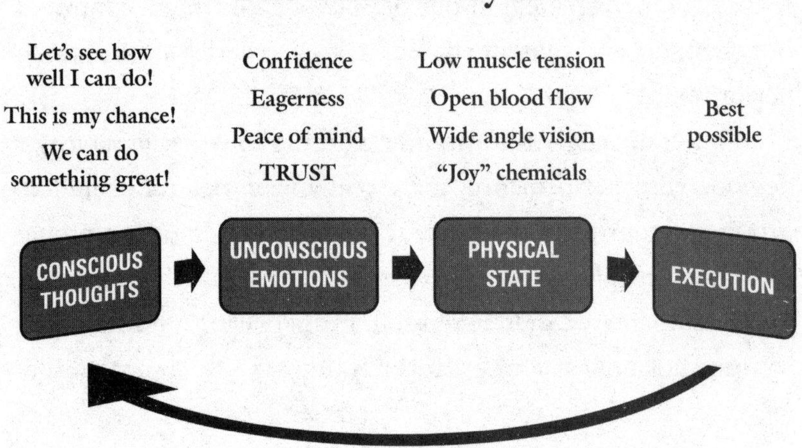

opens up instead of closing in; there are natural pain-reducing chemicals released in the brain. All these changes make your best possible performance far more likely, as pictured below:

Three key points about these cycles are important to winning the First Victory. First, we all find ourselves switching back and forth from the not-so-helpful sewer cycle to the more helpful success cycle many times a day and even many times an hour. No one, not the mentally toughest, most confident individual on the planet, is immune to the occasional trip around the sewer cycle.

What matters, then, is (1) how often and how frequently you're on which cycle and (2) which cycle you're on when it's time to perform. Take this simple two-question test right now. Answer honestly: What percentage of your conscious thoughts over a day, a week, a semester, a season, or a year are personally affirmative and emotionally supportive, and what percentage are personally disparaging and emotionally discouraging? And where do your thoughts tend to run when you are moments away from stepping into your personal performance arena? Do they move you toward the sewer or more toward success? Your answers suggest whether you are enabling or inhibiting a First Victory.

Second, no matter how you answered these two questions, the good news is that you have a choice point in your thought/performance cycle. No matter the outcome of any particular performance, you can deliberately choose to linger on the proper thoughts about it as well as the proper thoughts about yourself. No matter what the quality of your execution has been and no matter how long you have been maintaining a certain ratio of helpful thoughts to hurtful thoughts, you can choose to *change your mind* and get yourself on that constructive cycle more and more frequently. When I first started working with Eli Manning in early 2007, he was honest enough to admit that his thoughts about his play were 50/50 "good and not so good." That ratio changed dramatically over the next ten months, contributing to Eli's first Super Bowl victory and MVP award. How you can change your ratio and remain more consistently helpful will be described in the following chapters.

Third, being more consistently helpful in your thinking is not a guarantee that you will perform magnificently or win every game you play. Winning the First Victory gives you the best possible chance of winning subsequent victories over external opponents. There's an old military saying worth remembering here: *the enemy gets a vote,* meaning you can do everything right in terms of your own preparation and execution, but the outcome of the game, the test, or the battle will also be influenced by what the "enemy" (i.e., the opponent, the competition, the customer) does. The one guarantee that I can make is that if you live, work, and perform predominantly in the sewer cycle, your performance will always be less than it might have been and the enemy's vote will be that much more decisive in determining any outcome. Maximize your chances for success by choosing to apply this mind/body/performance connection. As my longtime karate teacher, Tsutomu Ohshima, once put it, "Mentally stronger person always has better chance to win." So put yourself in the best possible state to perform by *creating an attitude of energetic curiosity.* You do this by deliberately thinking, *Let's see how well I can run/throw/sing/speak/study/listen right now!* instead of being dead serious and thinking, *This race/game/audition/speech/meeting is super important, so I really have to do well right now or I'm in big trouble!*

Pillar #2: Inevitable Human Imperfection. You Won't Escape It, So You Might as Well Make Friends with It.

Natalie Portman won the 2010 Best Actress Oscar for her role in the movie *Black Swan*, portraying a ballerina preparing to star in the New York City Ballet Company's production of Tchaikovsky's Swan Lake. Ruthlessly perfectionistic and driven by her childhood dreams of dancing in this role, Portman's character, Nina, descends into a self-destructive hell. As movie historian Jadranka

Skorin-Kapov put it, "The film can be perceived as a visual representation of Nina's psychic odyssey toward achieving artistic perfection and of the price to be paid for it." And in this movie the price is high indeed—Nina's single-minded drive to be perfect in every step, turn, and leap destroys her self-esteem and her ability to enjoy life. As she continues to train fanatically for this performance, she loses her grip on reality and experiences a series of bizarre hallucinations that culminate in bloodshed on opening night.

Okay, it's a movie plot and not real life, but there are disturbingly similar stories of talented and dedicated people ruining their careers because they cannot accept the simple truth that they will never be perfect. Kate Fagan's 2017 book, *What Made Maddy Run: The Secret Struggles and Tragic Death of an All-American Teen*, is a moving story of a tragedy resulting from collegiate athlete Maddy Holleran's destructive perfectionism. And if you've spent any time around serious ballet dancers, as I have, they'll tell you the *Black Swan* movie narrative isn't all that far from the truth.

The common thread in stories like these, real or imagined, is ambition and the drive to accomplish big dreams taken too far. What started out as useful, beneficial, and indeed necessary motivation to practice, study, and so on becomes destructive perfectionism, the compulsive striving toward impossibly high standards, and the simultaneous self-criticism and negative judgment of oneself whenever that standard isn't achieved. Nothing will drain your confidence account faster or prevent your First Victory more surely than your refusal to accept and work with inevitable human imperfection. If you punish yourself for each and every mistake, error, and imperfection, it's impossible to be confident.

Destructive perfectionism is different from *striving for perfection*, the discipline and dedication necessary for sustaining your efforts toward improvement. A small amount of perfectionism is absolutely required if you're going to develop your knowledge, skills, and fitness, just as a small amount of spice is required in most cooking recipes to give the meal some excitement. But too much perfectionism will derail your progress and ruin your life, just as too much spice ruins any meal.

Every human being, no matter how talented or accomplished, is physically, technically, and mentally imperfect. That means you, your boss, your collaborators, and your competition are all going to make mistakes now and then. No matter how long you practice, no matter how hard you study, and no matter how careful you are, you will never be perfect in your sport, at your job, or in any other role you play (spouse, parent, sibling).

If the idea of never becoming perfect at something you care deeply about bothers you, let me offer a little reassurance: you can pursue your chosen craft or profession with fiery passion, and you can be very successful, very inspirational, even the best in the world at it without ever being obsessed with the need to be perfect. Research shows that the highest achievers in any given field are those with only moderate levels of perfectionism. And those with the highest levels of perfectionism are only moderate achievers, because the anxiety they feel over making mistakes prevents them from taking timely action.

So how do you get it right? How do you take advantage of perfectionism's healthy energizing qualities without letting it become destructive? Here are a couple of important guidelines I share with hundreds of cadets every year at West Point:

Strive for perfection, but don't demand it. Understand that you're not going to achieve perfection, but go for it anyway. Attack each task, each point, play, heat, stroke, and each meeting with a "Let's see how great I can do this" attitude. Maybe you'll perform it beautifully, achingly close to perfect, and maybe you won't. If you do, great! Enjoy the moment. But when you don't achieve that desired level of perfection (which will often be the case), don't beat yourself up as some kind of loser. Instead, look objectively at what you can do differently in the future, tell yourself that that's what you will do next time, and then forget the imperfection ever happened (more on this in Chapter Three). It's the negative reaction you have to your human imperfection and not the imperfection itself that drains your mental bank account and prevents your First Victory. Again, the science on this is conclusive. To quote one published study on the relationship between perfectionism and anxiety: "Those athletes who strive for perfection while successfully controlling their negative reactions to imperfection experience less anxiety and more self-confidence during competitions."

Be curious about your imperfections. They are valuable sources of information. You can actually gain a measure of confidence from each mistake, setback, or imperfection, and that is exactly what confident people do. They view any imperfection from a somewhat detached angle. With minimal emotion, they ask themselves, *What is this mistake telling me?* and then, *What will I do differently next time to make it turn out better?* It's this sense of curiosity about their imperfections that keeps them learning and growing. Instead of producing frustration and irritation, imperfections viewed in this way become friendly stepping-stones to success. If you're going to make mistakes (and you will), you might as well benefit from them.

One excellent example of functional perfectionism is Greg

Louganis, winner of five Olympic medals (four golds and a silver) in springboard and platform diving in the 1980s. Louganis is a self-described perfectionist, "but that's the irony," he says. "In order to do it perfectly, I have to let go of perfectionism a little. In diving, there's a sweet spot on the board. I can't always hit it perfectly. Sometimes I'm back from it, sometimes I'm a little over. But the judges can't tell that. I have to deal with whatever takeoff I've been given. I can't leave my mind on the board. I have to be relaxed enough to clue into the memory tape of how to do it. That's why I train so hard. Not just to do it, but to do it right from all the wrong places."

I would love to ask Greg Louganis just how often he did hit that "sweet spot" during Olympic and other world-class competitions. I'd bet it was a low percentage of the time. I'd further bet that for most of the dives that earned him his gold medals, he did not hit the sweet spot perfectly but then "let go of perfectionism a little" to produce a beautiful dive and earn the top scores. Knowing he won't hit that perfect spot on the board every time but refusing to let that imperfection affect the rest of his dive (the takeoff, the lift, the execution, and the entry) is a crucial element to Louganis's success in a sport where the thinnest margins separate the winners from the also-rans. The slightest bit of regret over not hitting the board's sweet spot, the tiniest sense of frustration resulting from not getting it just right, will create enough tension in his body to produce a pronounced effect on his subsequent execution and his score. Being committed to his success as he is, Greg Louganis refuses to allow that regret and frustration to take precedence. He strives to hit that perfect spot every time, but then accepts whatever he gets and stays "relaxed enough" (no worries) to let a great dive happen.

What is your typical response when you don't hit the sweet spot in your work? Do you relax into the next moment or do you tighten up? Accepting your human imperfection will help!

Pillar #3: Your Helpful, But Mostly Misunderstood, Autonomic Nervous System. Fall in Love with Your Butterflies.

Here's part of a conversation I might have with you during our first or second meeting. I've had versions of it a thousand times, with players from every conceivable sport and with performers from the worlds of medicine, business, and the performing arts:

YOU: Doc, I'm really good at practice every day, but as soon as I get to a game I kinda get psyched out. I get real nervous and real tight and my mind starts going a million miles an hour.

ME: How do you know you're nervous? Tell me what is going on that tells you, "Hey, I'm nervous."

YOU: Well, I can feel my heart speeding up and my palms getting sweaty and my hands getting jittery, and then my stomach acts up, flipping around like crazy.

ME: Okay, got it. So once you notice these things going on, what goes through your head?

YOU: I get all anxious, really uncomfortable. Like I said before, my mind starts going a million miles an hour.

ME: And when your mind starts going fast like that, is it a rapid-fire sequence of thoughts about you doing really well in the game that's about to start?

YOU: No, no, no, Doc. I told you, I'm really nervous. I start worrying like crazy.

Let's hit pause for a moment and analyze the situation. You're about to step into the spotlight for a performance and your body seems to be going into some kind of hyperdrive—a racing heart, twitching muscles, sweating palms, and those well-known stomach butterflies. These physical sensations tell you that you're "nervous" and being "nervous," in your mind, is a cause for worry. Welcome to the greatest "psych-out" in the world of human performance—the misunderstanding and misinterpretation of the human body's natural and beneficial process of physiological arousal. What you've told me is that the arousal you feel in the form of a racing heart, twitching muscles, sweating palms, and stomach butterflies is a signal that something is wrong with you. What you need to know is that this arousal, a natural process of your autonomic nervous system, is actually your friend and ally. That arousal shows up for the sole purpose of taking your performance up a new level. Here's how it works:

The dictionary tells us that *nervous* can mean either "easily agitated or alarmed" or "relating to the nerves." The latter definition is a lot more helpful. Being "nervous," as far as I'm concerned, simply means your nervous system is more active—the neurons in the brain, spinal cord, and throughout the periphery of the body are all buzzing faster and brighter than usual. Why would the nervous system ramp up like that? Simple: whenever you're about to do something that matters to you, whether it's something you MUST do or something you WANT to do, your autonomic nervous system—the part of your biology that keeps your heart pumping, your lungs breathing, and your digestion all working without any conscious effort on your part—does a few things to help you out.

The same way a general orders down to his soldiers through a chain of command when it's time for mobilization, the unconscious part of your brain that knows you're about to perform sends signals flying out to every part of the body, telling organs, muscles, and glands, *Hey, something important is about to happen. All units report in!* One place where these signals go is to the adrenal glands, two small wads of tissue that sit atop your kidneys. Being good little soldiers, the adrenal glands do what they're told when they get the signal; they report in and perform their one and only function—dumping adrenaline into the bloodstream (it might be a little adrenaline, it might be a lot, all depending on how much your brain thinks the situation requires).

That adrenaline finds its way back to the heart through the magic of your circulatory system, and from there it gets distributed throughout your body, everywhere the blood goes. Wherever it goes, things start getting active: the heart muscle itself, infused with some "adrenalized" blood, pumps harder (and hence loud enough for you to really notice it); other muscles throughout the body receive their share of the enhanced blood and, together with more "prepare to fire" neuron signals from the brain, twitch in anticipation, making for the jitters you feel in your hands. And the one hundred million neurons connecting the brain to the stomach and intestines also fire faster, making the sensitive smooth muscle fibers in the stomach vibrate like butterfly wings.

The end result of all this activity is that you become a stronger, faster, more alert, more perceptive (your pupils open up, too), more fully prepared to take on the world human being! Essentially, your own body, all on its own, without any conscious effort, produces a state-of-the-art, custom-made, performance-enhancing chemical for your unique biochemical needs and delivers it in precisely the right dosage at precisely the right time when it can do you the most good. And it doesn't cost you a dime. And unlike many other performance-enhancing chemicals, this one's perfectly legal! Stop for a second and think about how truly wonderful it is that your body gives you such a powerful gift when it senses you could use a little help.

This gift of adrenaline and sped-up neural activity also produces a few unwanted side effects, and these side effects account for a lot of confusion. That racing heart, those jittery muscles, and butterflies in our stomach—the very sensations that you identify as the reasons why your "mind starts going a million miles an hour," the very things that tell you that you are "nervous"—are in fact signals from your body that it has tapped into some high-octane rocket fuel and is now ready to perform. If you are experiencing any of these signals, it simply means your body is doing something to help you do something that matters to you.

At this moment of truth for your First Victory, then, what will you *think* when the rocket fuel kicks in? Remember Pillar #1—your thoughts drive everything. Will you think, *Something natural and marvelous is happening to help me be great, so let's see how great I can be* and engage the success cycle? Or will you think, *Uh-oh, I'm freaking out, this is really bad* and fall back into the sewer?

Unfortunately, I have seen far too many performers choose the damaging sewer option. Why? It probably has to do with your early experiences in performance moments, when you were new to a sport or a task or a situation that you hadn't yet developed skill or competency at. Being unskilled, it's likely you experienced relatively little success, and as each of these unsuccessful moments were preceded by your body's natural adrenaline and arousal, you likely learned to associate this arousal with a soon-to-follow disappointment. This association can stick, even after you've practiced/studied enough to attain high levels of skill and competence; you still have the sense that something is about to go wrong when you they experience the butterflies.

That association need not stick forever. Anyone can "change the narrative," as *Start with Why* author Simon Sinek points out. In one of his "Simon Says" videos, Sinek shares this advice for reframing nervousness into excitement: he points out that when you're "nervous," your heart races and you envision a future (usually a bad one), but that when you're "excited," your heart also races and you also envision a future (usually a pleasant one). Sinek gets the key point right—the underlying biology of both nervousness and excitement is the same: the naturally occurring arousal that evolution hardwired into human physiology, a legacy from our primitive ancestors, who benefited from being able to mobilize energy quickly to deal with the uncertainties of prehistoric life. How we interpret that arousal, the "narrative" we tell ourselves about it, determines whether we feel uncomfortably "nervous" or functionally (maybe even pleasantly) "excited." We can choose to interpret that arousal as either beneficial or damaging, as either a blessing or a curse.

I hope the changing of your personal narrative about arousal, your reinterpretation of that helpful autonomic nervous system activity, has begun upon reading this section. It can happen for you in a single moment, just as it did for former NFL wide receiver Hines Ward. As reported in *USA Today* the day after the 2006 Super Bowl, Ward changed his narrative about being nervous when he got some reassurance from another veteran player that it was perfectly okay for him to feel some stomach discomfort prior to such a big moment. "Ward then headed to the bathroom, took care of his upset stomach, and went on to make five catches for 123 yards en route to being named the game's MVP."

However it happens for you, the conclusion that matters for anyone seeking to win the First Victory of confidence is the conclusion expressed by Olympic sprinter Michael Johnson right after he made history as the only man to win both the 200 m and 400 m sprints in the same Olympic Games in 1996. Interviewed by NBC's Bob Costas after the Games, Johnson was asked if his heart was pounding as he stepped into the blocks for the 200 m final, having already won the 400 days earlier. Johnson replied, "Definitely, my heart was pounding. I was nervous." Then he added, "and when I'm nervous I'm comfortable." Pause for a moment and digest that statement: *when I'm nervous I'm comfortable.* Rather different from how most people (even many veteran performers) view being nervous. Johnson's statement shows the degree to which he has changed the common narrative of nervousness from a state of dis-ease into a source of power, to something he actually looks forward to. By changing his interpretation of nervousness from foe to ally, Johnson wins another First Victory.

Back to the conversation I'm having with you.

ME: Now that you know this, how about accepting the fact that being nervous really means you're getting ready to be at your best? How about changing your narrative about it and deciding to be *comfortable* when you're nervous?

YOU: I never thought about it that way. It makes so much sense.

But then the old misunderstanding typically comes back in two ways.

YOU: But, Doc. It feels so *different* when I'm nervous like that. It doesn't feel *normal* at all.

ME: Of course it doesn't *feel normal*. Why should it? You're about to do something that matters more to you than some meaningless *normal* activity like filling up your car with gas or brushing your teeth before bed. Why on earth would you expect to feel *normal* about it? Champions like Michael Johnson know that it won't feel normal when they step into the spotlight, and they look forward to that very feeling as a signal that something special is about to happen.

YOU: That's certainly different from what I was always told. I always heard about the best being "cool under pressure" or "having ice water in their veins."

ME: Nothing could be further from the truth. Their blood is just as hot as anyone else's, but they appear cool on the outside because they've learned to:

1. Respect their autonomic nervous system's intelligence
2. Expect their nerves to fire up before they start something important
3. Embrace their newly produced energy.

Once that point is digested, the old misunderstanding takes one more final shot.

YOU: I've been doing this sport for years and I'm pretty good at it. I figure by now I wouldn't need to be nervous before games.

ME: Your nervousness (though I hope by now you have changed the narrative to "excitement") is the result of a process that became hardwired into human biology some two hundred thousand years ago, when the ability to mobilize energy during important moments (such as hunting or running from trouble) meant a better chance of survival. And even though we no longer depend on this primitive fight-or-flight response for survival, that ancient biological wiring still exists within each of us, and it will continue to operate no matter how experienced or competent you are.

New England Patriots head coach Bill Belichick, even after coaching in the NFL for forty-four years and through six Super Bowls, admitted in a January 2019 TV interview that he still "gets nervous" before every game. "You want to go out there and do well. We all have things in the game that we have to do. You want to perform them well and not let your team down because everyone is counting on you to do your job." Certainly Belichick is as experienced and competent as anyone in his field, but the primitive energy mobilization process hardwired into his biology still fires up every Sunday. Thirty-year Naval Special Warfare veteran Richard Marcinko agrees: "Before an Op, everyone feels a certain amount of gut-wrenching, sphincter-puckering nervousness. I don't care how seasoned, proficient, or competent you may be, how cool you are under fire, or how many times you've gone shooting and looting. Until you're actually over the rail and the bullets are flying, you're going to experience a few butterflies."

Accepting this simple fact of human existence removes one more source of potential doubt and worry. Knowing this, you can be comfortable when you are nervous, and thus more certain of yourself in important situations. Respect your autonomic nervous system's intelligence, expect your nerves to fire up before you start anything important, and embrace the newly produced energy. These are important steps toward your First Victory.

Pillar #4: The Inconsistent and Delayed Returns of Practice. Great Changes Are Happening That You Can't See.

You've probably heard of the ten-thousand-hour rule—the idea that you have to devote ten thousand hours of practice if you expect to be an expert at your sport, instrument, or profession. And you may have seen the recent findings of the expert performance literature, telling you that it's not just any ten thousand hours of practice, but it's "deliberate practice"—that is, practice organized around specific guidelines—that produces the expertise we seek. The assumption that runs through all of these assertions is that steady, quality practice produces steady, quality results.

But two different realities of practice and improvement that have a huge impact on our pursuit of success and excellence have been left out of this narrative. First, the return on our "investment" of practice will be uneven and inconsistent at best;

no matter how diligently we follow every deliberate practice guideline, we will experience dry spells, long plateaus where it seems like we aren't improving at all. A plateau phase will be interrupted by a burst of improvement, followed by another plateau, followed by another burst and on and on and on. Nobody told us this. Labor does indeed have its rewards, but they are far from the sure thing we had been taught to believe they were, and hanging in there, putting in the work during all those plateaus when it seemed like we weren't improving at all, is a huge exercise in patience. Very little in our culture of instant gratification, twenty-four/seven access to a world of information, and immediate contact with anyone and everyone prepares us to work that patiently. No surprise that a lot of people give up on their dreams when they discover that the path to attain them is not one of continuous improvement but is instead rocky and uncertain.

Second, the longer we pursue achievement in our chosen sphere, the further we advance down the road to success, the longer those plateaus last before we feel an improvement happening, and the smaller those bursts of improvements become. Not only are the returns on our investment of work unpredictable, they also diminish the longer we make them. This reality is fertile ground for frustration and self-doubt. What's the point of working so hard if the payoff is not just inconsistent but also lessens over time? What if those long plateaus and those short bursts of improvement just prove that you don't have what it takes to succeed at that sport or that instrument or that profession? Maybe you should quit and try something else. First Victory lost.

Take heart—your investment of "work" is far from pointless, and those long plateaus and short bursts do not mean that you can't eventually succeed. Everyone else pursuing success—your peers, your competitors, and your opponents—are all experiencing these same realities. If you can accept them and work with them just a little better than they do, you'll create an advantage for yourself.

The way to minimize that frustration and set the stage for your First Victory is to understand that every minute of quality practice, every rep, drill, and practice session properly conducted, creates beneficial changes in your nervous system that ultimately, over time, bring about substantial improvements. Each of these changes is small, but they add up, and once they reach a certain critical mass they result in a noticeable "aha" moment: your tennis serve suddenly becomes more accurate, your fluency with French more automatic, your sales pitch to clients more authentic. We don't notice these improvements as we are practicing and studying, but the important fact, the important reality, is that they are happening all the time while we seem to be on the plateau treading water and feeling like we are getting nowhere.

In his 1991 book *Mastery: The Keys to Success and Long-Term Fulfillment*, the American educator and philosopher George Leonard described this delayed effect of practice in terms of an exchange between the brain's habitual behavior system and its cognitive and effort systems. Simply put, you have to apply conscious deliberate effort to create a new habit or change an old one, to learn how to hold a hockey stick as a beginner or refine your wrist shot once you're a veteran. Once your cognitive and effort systems have reprogrammed your habitual system through deliberate practice, the effort brain can step back and withdraw. Now you can hold that stick or make that faster shot without thinking about how you do it. "At this point, there's an apparent spurt of learning," Leonard wrote, *"but this learning has been going on all along"* (Leonard's italics). Knowing that learning has been going on all along will help you remain focused and upbeat while on one of those many plateaus. The changes in your nervous system, the processes that produce the desired improvements, take place while you are *on the plateau*. With this understanding, plateaus need not be dreaded or merely tolerated, but instead should be valued. Just as you can value the precompetition jitters as evidence of a beneficial energy boost, plateaus can be valued as your own "improvement factories."

Another useful insight into the delayed effects of practice is provided by recent neuroanatomy research demonstrating that the outer sheath of a neuron, the individual nerve cells of which our entire nervous system is composed, grows each time that neuron is activated. Each and every thought, feeling, and action we experience as human beings is brought about by a particular electrical signal traveling through a chain of neurons arranged in a complex circuitry. Right now as you read this page a series of sophisticated electrochemical reactions are sending impulses from the optic nerves in the back of your eyes along a highway of connected nerve cells to the visual cortex located in your brain's occipital lobes. When you play a piano scale or throw a ball or analyze a sales report, other equally complex neural pathways are activated. These circuits are composed of thousands of neurons that take in relevant information from our sensory organs, then link with thousands more to determine a response based on our memories and experiences, and then link with thousands more to trigger an action. Covering these nerve fibers is the substance myelin, a phospholipid (nerdspeak for a type of fat) that is produced in the brain and the spinal cord and acts like the insulation covering the copper electrical wiring in your home. Just as thicker layers of protective electrical tape or insulation allow an electric current to travel more rapidly down a copper wire, thicker layers of myelin, the human nervous system's naturally produced "insulation," allow electrochemical impulses to travel faster down a given circuit. As journalist Dan Coyle put it in the book *The Talent Code*, "myelin serves as the insulation that wraps these nerve fibers and increases their signal strength, speed, and accuracy. The more we fire a particular circuit, the more myelin optimizes that circuit, and the stronger, faster, and more fluent our movements and thoughts become." The more the circuit is fired, the more myelin is produced; and the more myelin produced, the more efficient the circuit becomes.

Viewed this way, improvements in human performance, whether it's shooting a basketball, solving a calculus problem, or delivering a closing argument in a courtroom, come about when the neurons that control that performance achieve a new level of efficiency, when the electrical signals passing through them move faster, smoother, and with more precise timing than before. Once that layer of myelin insulation becomes thick enough, as a result of repeated firings of the neural circuit, the speed of the impulse passing through that circuit increases up to one hundred-fold.

But here's the hitch—the building up of the myelin insulation sheath is slow. This process takes time. The circuits controlling the law student's understanding of contracts or the quarterback's understanding of next week's opponent have to be fired over and over again before enough myelin is laid down to optimize them, and that requires passionate, persistent practice. NFL coaching legend Vince Lombardi may not have known why "the dictionary is the only place where success comes before work," but he was right. Practice produces changes, but those changes happen slowly and unpredictably.

The upshot of Leonard's explanation of learning systems and neuroscience's identification of how myelin both functions and develops is that practice, particularly deliberate practice right at the edge of one's present ability, creates minor, immediate changes that are not always noticeable but that chain together over time to produce major, noticeable changes once a certain threshold is reached. The chaining, the building, the growing, the actual development we seek through practice happens while we are on the plateau itself, not during the bursts of improvement. What we have come to value most in our present-day world of immediate gratification is the breakthrough experience, that moment when all those small, imperceptible changes reach a critical mass and explode into a palpable advance. But true development happens on the plateau. Leonard concludes: "To love the plateau is to love the eternal now, to enjoy the inevitable spurts of progress and the fruits of accomplishment, then serenely to accept the new plateau that waits just beyond them. To love the plateau is to love what is most essential and enduring in your life."

If you are like most of the serious competitors I know, you probably don't actually "love" being on a plateau. But I hope that you, like those other competitors, develop an acceptance of the plateau, an understanding that as long as you're on it while putting down the proverbial hammer of quality practice, a number of wonderful but as-yet-unseen changes are underway. You don't have to understand any more about brain reorganization or myelin optimization—just know, deep down, that good things are happening, and those good things will surface and become obvious in time. Once you understand and accept this concept, you will establish the foundation of your First Victory.

CHAPTER TWO

Building Your Bank Account #1: Filtering Your Past for Valuable Deposits

Back in the 1990s, before the coming of video streaming services like Netflix, a common family ritual was the weekly trip to the local video store. There Mom and/or Dad could find a drama or action movie, and the kids could find a G- or PG-rated movie of their choice. Being dutiful parents, my wife and I embarked on many such trips, bringing home such classics as *First Wives Club*, *Wayne's World*, and *Mrs. Doubtfire*, (yes, I know I'm dating myself). We'd all huddle on the living room couch and crank up the now obsolete VCR machine. On one of these trips my two young daughters ran over to me after making their selection from the rack. "Daddy, can we get this one?" It was the recently released comedy *Dumb and Dumber*, starring Jim Carrey and Jeff Daniels as two bumbling adults with the common sense and worldly experience of ten-year-olds. I remember not being impressed with my daughters' selection, but being the doting dad that I was, I relented and prepared myself for a very forgettable evening. I was getting pretty much what I expected (slapstick gags and scatological jokes) when there appeared on my TV a scene that I had to stop, rewind, and watch again. It's the scene (maybe some of you have seen it) where Jim Carrey's character, Lloyd, a gangly, homely, and altogether unsophisticated man, has finally tracked down the beautiful Mary (played by Carrey's soon-to-be wife, Lauren Holly) and asks her if there is any way the two of them could ever end up together as a romantic couple. Not wanting to be rude, but utterly uninterested in Lloyd, Mary tries to let him down gently with a few noncommittal answers. Finally Lloyd insists that she give him a straight answer about what his chances are. "Not good," she says. "Like one out of a hundred?" Lloyd asks. "More like one out of a million," she answers, knowing that she has just broken the poor man's heart. Upon hearing this pronouncement Lloyd at first swallows hard and bites his lip, disappointed that his quest for love seems hopeless. But after a moment's reflection he smiles a huge, gap-toothed smile and proclaims, "So you're telling me there's a chance!" before breaking into a howl of glee. He has determined that he actually has a chance at love after all. Not much of a chance, to be sure, but enough of a chance *in his own mind* to be a cause for celebration. First Victory won.

In that moment Lloyd exhibited the single most important mental skill anyone can have, and indeed must have, to build confidence in an indifferent and often uncaring world. He is thinking *selectively*—only allowing into his mind thoughts and memories that create energy, optimism, and enthusiasm. He may only have one chance in a million, but he is completely focused on that one chance and consequently he feels on top of the world. Because of the ongoing mind-body connection presented in Chapter One, Lloyd's optimistic feeling will keep him pressing on in his persistent pursuit of love.

Put differently, Lloyd displays a remarkably effective *mental filter*—a screen through which all his thoughts and experiences pass before they become part of his running total, before they can affect his mental bank account—either building it up or bringing it down. This filter performs two functions in the service of that mental bank account. It allows the thoughts and memories that create energy, optimism, and enthusiasm to pass through and build the mental bank account up, but it prevents the thoughts and memories that create fear, doubt, and worry from getting in and drawing the bank account down. With a functioning mental filter, you could play in an afternoon softball game, get only one hit in four at bats, and spend the evening reliving and enjoying that one successful trip to the plate. Naturally, you might also want to sharpen up your batting stance and your swing with a little practice before your next game, but by hanging on to the memory of that one hit, rather than beating yourself up by reliving the three misses, you give the part of your nervous system that sees the ball and swings the bat an image of what you want more of, and thus you get that mind-body connection described in Chapter One working for you.

This is precisely how the best hitters in baseball have always thought. Hall of Famer Tony Gwynn exercised his mental filter by editing the video of each at bat after every game. Into one video file went each pitch where he made solid contact with the ball. Into a second file went each pitch where he made a good decision to either swing at the right pitch or hold back from swinging at a bad pitch. Into a third file went all his bad decisions, either holding back on a good pitch or going after a bad one. That third file was promptly and permanently deleted. Why throw it away? "The last thing I need to do," said Gwynn, "is watch myself looking like a fool swinging at somebody's curveball." Great performers in all walks of life have always had powerful, effective personal mental filters. Regardless of what actually happens to them, they perceive, through their filters, all their experiences in the world in ways that help them move toward success. When they are successful, even in minor matters (e.g., success in a particular drill during practice, a good grade on a minor quiz or report), they focus completely on this momentary success, allow themselves to feel skilled and proud because of it, and assume that the success will happen again. Their filters permit constructive experiences, no matter how small, to readily pass through and become a permanent deposit into their mental bank account. When they are less than successful, they either release the memory completely or restructure it so that has no negative effect on their confidence.

For an inspirational example of how a mental filter can work effectively even in the most serious of circumstances, the story of retired U.S. Army Captain John Fernandez is hard to beat. A likable, hardworking, blue-collar kid from Long Island, John Fernandez graduated from West Point in 2001, having been elected captain of the men's lacrosse team his senior year, despite not being one of the team's all-Americans or high-scoring stars. In April 2003, now Lieutenant Fernandez was heading north from Kuwait toward Baghdad with his Field Artillery platoon. Operation Iraqi Freedom was underway. The pace of the army's advance was fast; after two days with no sleep and the column halted for the night south of Baghdad, John decided to catch a nap atop one of the convoy's Humvee vehicles. Little did he suspect that he was about to become part of one of the early tragedies of the Iraq War.

Unbeknownst to John as he settled into his sleeping bag, flying overhead was a US Air Force A-10 Thunderbolt. Mistaking the movement around John Fernandez's resting column for enemy activity, the pilot of the A-10 released a five-hundred-pound, laser-guided bomb that detonated on impact close enough to where John lay sleeping to send him flying to the ground and severely damaging both his legs. Had he been sleeping in the opposite orientation, with his head toward the blast instead of his feet, he would certainly have been killed. Waking up hours later in a field hospital after being medevaced to safety, Lieutenant Fernandez received the grim news that two of his soldiers had died in the blast, and that both his legs, the right one from just below the knee, and the left one from the lower calf, would have to be amputated. The vigorous young man who had been a collegiate athlete would spend the rest of his life in a wheelchair or walking on artificial limbs. "I decided right there that I would never feel sorry for myself," John told me years later. "I told myself that I was going to have a great life." I have heard John tell cadets several times, when he has returned to West Point as a guest speaker or to take in a lacrosse game with his wife and children, "It's really no big deal . . . You get up in the morning and put on your shoes and socks. I get up in the morning and put on my feet." That, ladies and gentlemen, is an effective filter.

This chapter is devoted to the construction and use of that filter, so that you make the maximum number of deposits every day and thus build your confidence. *Construction*, in fact, may not be the right term. You don't have to create a personal mental filter from scratch, because *you already have one.* In fact, *you've always had one and it's operating right now.* In this moment your mind is "letting in" certain elements from both the outside world and the internal world of your thoughts, and "screening out" certain others. It's busy every waking moment interpreting the world around you, recalling both recent and long-ago memories, and conducting a nonstop symphony (or shouting match) of internal chatter. How you "filter" all that mental activity, what parts of it you pay attention to and what parts you choose to ignore (like Tony Gwynn did), will determine how you feel about yourself and whether you win the First Victory. Like Tony Gwynn, you're creating video files from all your experiences every day. The only question is whether your filter is working for you to build up your mental bank account of constructive thoughts and memories or whether it's keeping thoughts of effort, success, and progress out of your mind and thus working against you.

If your filter isn't creating a lot of excitement for your future the way Lloyd's does for him, it's probably because you've been led to believe that keeping your mind full of memories and thoughts that produce a lot of energy and enthusiasm is somehow unrealistic or improper, that it might be okay for someone else but certainly not you. That belief is in fact making it harder for you to be good in your chosen field. That belief simply encourages you to stay fixated on your shortcomings, failures, and imperfections, the very opposite of what you want. Think about it. If you're not remembering your accomplishments, dwelling on your strengths, and envisioning a desired future, then just *what are you remembering, dwelling on, and envisioning?* Setbacks and disappointments, most likely. The science on this is pretty clear: whatever your conscious mind tends to think about most is what your unconscious understands and tends to work toward. Consequently, that's what you tend to get more of.

Fortunately, it doesn't have to stay that way. One of the best features of your mental filter is that you have control over it. You can choose to let in the thoughts and memories that create energy, optimism, and enthusiasm or those that create fear, doubt, and worry. We all have a human ability called free will—the ability to choose the thoughts that make up our waking consciousness every passing minute. It's that ability "to choose one's attitude in any given set of circumstances" that psychologist Viktor Frankl identified as the "last of the human freedoms." You can choose to filter in the constructive aspects of what is happening every day in your life, and thus build yourself up, or filter in the negatives and drag yourself down. There is really is no middle ground here, and it is where the First Victory begins. Everything else can be taken from you, but this power to choose what you think, what you remember, and what you believe about yourself is untouchable. You have it now and you will have it always.

Your mental filter operates on three levels. It filters the memories of your past, everything from long ago to what just happened yesterday and what happened five minutes ago. It filters your thoughts about who you are right now and what you are capable of. And it filters how you imagine your future, what you will do and how you will do it. The unifying thread across these three dimensions is this simple principle—the discipline to think about performing well and ultimately succeeding whenever you think about your chosen field. If you are a car salesman, you think about making great deals and providing a great experience for your customers every time you think about your job. If you are a medical student, you think about mastering the material in each of your courses and having a great career every time you think about your work. If you are a tennis player, you think about hitting great shots, winning matches, and having the season of your dreams every time you think about tennis. It's indeed a challenge to think this way about your chosen field, and no one ever succeeds at doing so 100 percent of the time. But the better you get at it, the more you keep that filter letting in the deposits and preventing the withdrawals, the more of an advantage you create for yourself.

Poor Lloyd in the movie lacks a backlog of useful experiences from which he could draw. He lacks the talents and skills that could help him right now. His future prospects are exceedingly dim, to the tune of "one in a million." But his remarkable filter keeps him firmly convinced that his one chance will come through. I am sure you, the reader of this chapter, have more useful experiences to draw from than Lloyd does. I am sure you have a ton more talent and a better support system right now than Lloyd does. And I'm sure you are facing far better odds than one in a million. But is your filter as effective as Lloyd's? Do you allow thoughts of past successes, present improvements, and future achievements to dominate your mind the way Lloyd allowed his one chance in a million to rule his mind and thus motivate him to persist in his quest for love? Imagine how powerful you might become if you combined your useful past experiences, your present resources, and your future prospects with Lloyd's remarkable filter. This chapter and the two that follow will teach you what you need to do to make that combination come about. The remainder of this chapter is devoted to techniques for managing your memories, filtering your past experiences so that you establish your mental bank account and start making consistent daily deposits. The following two chapters will present techniques for filtering your present thoughts about yourself and your visions of a desired future.

Mining the Memories of Your Past

We begin the process by filtering in some constructive memories. It's no secret that humans are motivated and influenced by their memories. The famous Sigmund Freud theorized that memories of our earliest childhood remain in our unconscious and control us for as long as we live. While that theory has been debated fiercely, even the sternest anti-Freudians acknowledge that the subset of our memories which we hang on to with the greatest clarity and the greatest feeling influence our behavior in the present and our expectations of our future. That subset impels us either toward confidence and trust or toward doubt and worry.

Exercise One: Your Top Ten

Think back to when you first started participating in your chosen field (from here on in I'll use this term, *chosen field*, to refer to whatever it is that matters most to you and whatever it is that you desire to excel in—your chosen sport, profession, etc.). What did you enjoy about that activity? What made it cool, or fun, or interesting for you? It may be hard to put a finger on it but I'm sure there's a feeling, a rather special feeling, back there in your memory. Center yourself on that feeling and just sit with it for a moment. What's the "picture" that comes right along with that feeling, the still photo that pops up or the short video that plays in your mind as you feel it? Whatever it is, that's the first deposit into your mental bank account, the seed money that will grow into a personal fortune. I recommend you write it down in a notebook. You can write it into an electronic document or on a phone app if you like, but science tells us that using a pen/pencil and paper creates a stronger memory. Keep that notebook or electronic file handy, you're about to put a lot more into it.

Does that initial feeling and picture trigger a few others? If you're like most of my clients, both amateur and professional, a multitude of scenes from your past have been sitting in the back of your mind for a long time—scenes of pleasant, positive, probably exciting moments of you participating in your chosen field as a beginner or novice, and then progressing right on through to yesterday. Write these down too!

Now we're rolling. Each scene you identify, each memory you bring up, is a deposit into your mental bank account. You are now contributing to a running total of energizing and encouraging thoughts about your chosen field. This is what confidence is built on—this is the process of attaining your First Victory.

Time to take it up a notch. Let's mine your past for some forgotten gems and deposit those valuable memories into your bank account. This is an exercise I call the Top Ten. As the title suggests, it's about bringing the ten most-encouraging, most-energizing memories out of the dark recess of your mind and polishing up those jewels of thought so they radiate their brilliance back to you. Get out a blank piece of paper, put "My Top Ten" at the top of the page, and write out a list of ten accomplishments in your chosen field. If you're a competitive athlete write down the ten best moments playing your sport—games or races won, goals scored, and so on. If you're a musician, write down the most beautiful or memorable pieces you've ever performed, whether they took place on a large stage in front of a big audience or privately in your practice space. If you're a "white-collar athlete," one of the zillions of people who work daily to turn the wheels of the economy, compose a list of the projects you've completed, the clients you've served well, the contributions you've made to your organization's success. If you're a student, put down the papers you've gotten your best grades on, the compliments you've received from teachers, the exciting ideas or concepts you've learned. One young golfer I worked with started off his My Top Ten list this way:

1. flawless play in the North South Junior in '96 . . .
2. great play in the MGA Junior in '96 . . .
3. hitting all those greens in the '97 BC Open . . .
4. making that wedge out of the trees to 10 feet from the cup in the '98 Canon Cup . . .
5. turning it around on the 6th hole at the '99 Rockland Junior and finishing great . . .

Your list need not be awe-inspiring or jaw-dropping to serve as valuable deposits into your mental bank account. It doesn't matter if you haven't won a world championship or a Nobel Prize, or even your local tennis club's annual Fourth of July tournament. Whatever you accomplished in your own life qualifies. Every stay-at-home mom or dad who takes care of a preschool child and keeps the house clean has tons to be proud of, like teaching that toddler to say "please" and how to share her toys with the other kids at the local playground. Every law student, med student, art student, and auto mechanics student got to where they are now because they did some things, made some things, and mastered some knowledge that qualified them for more advanced study. Constructing this Top Ten list is the first exercise of many that you will be doing to develop the valuable skill of selective thinking. Once your list is made, attach a photo to it, either of you in action doing what you love or one of a meaningful accomplishment you are striving for. Here's an example of a collegiate wrestler's Top Ten list—from 2021 West Point graduate and NCAA Wrestling Tournament qualifier Bobby Heald. It shows a simple but effective format for bringing your Top Ten moments into clearer focus. At the top goes your name and the name of the team you are on right now or hope to join. Below that is the photo of you making it happen or of your immediate goal. Then list your Top

Ten moments, and at the bottom, write that goal down as a final reminder.

Put your Top Ten poster up on a wall where you'll see it often and be reminded of your accomplishments and progress.

I've had some clients balk initially at creating a Top Ten list because they think their past experiences don't mean anything in their present situation. These clients are often the ones who first declare to me how good they used to be in their chosen field and in the next breath tell me how they've lost all their confidence. Here's how the dialogue typically went:

PLAYER: Doc, I don't know what to do. I was a three-year starter in high school, and captain and MVP of my team senior year. I was the leading scorer junior and senior year and All-State both years. I was recruited by a dozen colleges, and last year I had the best stats of any freshman on the team. But I just don't feel confident anymore. Maybe I should just quit.

ME: Do you ever think back to those games in high school where you dominated, or to getting those All-State awards, or putting up the best numbers of any freshman last year?

PLAYER: Geez, I haven't thought about any of that for months. That was all a long time ago.

ME: Hmmm . . . seems like you've forgotten a lot about yourself, particularly about all the wonderful moments you've had playing your game really well. I think you're overlooking a real source of strength and comfort.

PLAYER: But all that's in the past. I'm at a whole new level of competition now and what I did back then doesn't matter anymore. Just because I was great back then at that level doesn't mean I'm gonna be good now at this level.

ME: So you're telling me you were a big fish in a small pond but now you're just a little fish in a bigger pond.

PLAYER: Exactly.

ME: Okay, I think you're a little confused when it comes to fishes and ponds. Let's say you were the biggest, strongest, healthiest fish in that pond back home. Now let's imagine that the local Fish and Wildlife Service came by with a net, scooped you out of that pond, and then released you into a much bigger pond, giving you a lot more room to swim around in and giving you a lot more to eat. Got that image? Good. Now tell me what happens to a healthy fish when it's put into a bigger pond with all that space and all that food?

PLAYER: It gets bigger, it grows.

ME: Exactly right! And that's what's happening to you right now, only you're not recognizing it. If you think that you've somehow gotten smaller upon being put into this bigger pond then you're in trouble. You're still the same big, strong, healthy fish you were back in that smaller pond. You need to remember just how big a fish you are!

PLAYER: Wow! I never realized I was undermining myself by thinking that way!

How big a fish are *you*? Put together your My Top Ten list and I bet you'll be pleasantly surprised by the answer.

Exercise Two: Your Daily E-S-P

Now that you've got your bank account started, the work of deliberately building it up and protecting it from losses really begins. The confidence that you want to feel six months from now when you take the MCAT or compete in that conference tournament is either built up day by day or allowed to diminish day by day. In fact, every day is a source of deposits, and you will find them if you choose to look for them.

If you're reading this toward the end of the day, reflect back on the events of the day. If you're reading this more toward the beginning of the day, reflect back on yesterday. In either case, your most recent practice session, training session, study session, workout, or workday presented you with an opportunity, probably several, to constructively contribute to the running total of thoughts that composes your confidence. Put that episode of your life through your mental filter. What took place during that episode, and what did you do during that episode that left you with (1) a feeling of pride in having given a good effort, (2) a feeling of accomplishment, and (3) a feeling of having made progress?

Take out that notebook where you started My Top Ten, open it to a fresh page, and at the top put in the date of the day you are filtering. Put a capital *E* next to the left-hand margin on the next line down. Now write down one instance of quality *Effort* from that episode (your practice, training, workout, workday). Identify one moment where you buckled down and honestly gave it your all. It could be that one drill in practice where you dialed yourself in. It could be that one station in the weight room or that one interval on the track where you were tempted to ease off a little but didn't. It could be that stack of papers that you forced yourself to organize and file or that last batch of email that you took care of before leaving for the day. Just answer the question "Where did I do some valuable work today?" Write down your moment of honest effort for the day (and if you have more than one, go ahead and enter them all).

Once you've completed this *Effort* entry skip a line and put a capital *S* next to the left-hand margin. Now write down one *Success* that you experienced during that same episode, one moment where you got something right. It doesn't have to be a big success. A goal scored while tightly covered, a sequence completed without a single form break, a PR set on the bench press all qualify. So does a report submitted on time, a compliment or thank-you received, or a positive number when the day's receipts are counted up. Just answer the question "What did I get right today?" Write down your *Success* of the day, however small it might be (and if you have more than one, you know what to do).

Once you've completed this *Success* entry skip a line and put a capital *P* next to the left-hand margin. Now write down one instance of *Progress* that you experienced during that same episode, one moment where you got better at something even if you didn't get it completely right. Did you get that backlog of requests whittled down? Did you get closer to running all the intervals at the designated pace? Did you improve your relationship with a coworker or get closer to yes in that negotiation? Just answer the question "What did I get better at as a result of my effort?" Write down your *Progress* of the day, however small it might be (and if you have more than one . . .).

This daily E-S-P (Effort–Success–Progress) reflection and

journaling process deposits at least three constructive memories into your bank account every day (and you could easily make ten such deposits depending on how thoroughly you reflect). It might take you five minutes to complete this exercise, but those five minutes will ensure that you constructively contribute something to the development of your confidence each and every day. Think of this as making a brief highlight video of your day's performance, just like the highlights you might see of soccer star Megan Rapinoe's World Cup goals or tennis star Roger Federer's winning shots on ESPN. Your daily successes and your daily progress might not make the evening news, but they are nonetheless the building blocks of your confidence and thus deserve your attention. By taking note of each day's highlights and *allowing yourself to feel good about them*, you subtly but powerfully encourage yourself to repeat the actions that produced that success and progress. The research in positive psychology tells us that when we experience those positive emotions like pride, excitement, and sense of accomplishment as a result of our reflections, we broaden our repertoire of actions and build resources that we can draw on in the future.

NHL goalie Anthony Stolarz used this daily reflection as an up-and-coming player, sending me a text message after every practice and game for an entire season. His messages were simple and straightforward ("stopped a two-on-one breakaway in the first period; made a great glove save on a shot from the slot; came right back and recovered my emotions after giving up a goal"), but they were valuable deposits in his mental bank account and helped him become the starter for his team and an American Hockey League All-Star (Stollie, as he is known, currently plays for the NHL Mighty Ducks of Anaheim).

Sometimes a client of mine will ask "What if I do my reflection and can't come up with anything? What if I can't find an E or an S or a P for each day?" I have a simple answer: you're not looking carefully enough. Go back and look at the workday or the workout/practice/lesson hour by hour or even minute by minute. Can you honestly say you didn't give honest effort anywhere at all, that you didn't get right at least one little thing, that you finished up absolutely no better at all at everything you did? Usually just a minute of head scratching is all it takes for even most cynical and negative person to find the right moments in their practice or in their day.

Consider for a moment what happens if you *don't* reflect on your experience and filter it. Yes, you'll probably keep moving and pushing along, but you'll miss out on many opportunities to build up your bank account and accumulate a storehouse of positive memories that you can draw on when it's time to perform. Just as there is the money management principle of "lost opportunity" where you lose out on the potential interest that a sum of money could earn for you over time by spending it before you have to, there is the mental management principle of "lost opportunity" as well. How many opportunities to win a First Victory are you missing out on by not taking advantage of your daily experience? Without meaning to, you are likely contributing to your own stagnation by letting moments of effort, success, and progress go by unacknowledged.

Exercise Three: The Immediate Progress Review, or IPR

This brings us to the last and potentially most powerful method of building your bank account by filtering memories from your past. Your account will grow fastest when you make the greatest number of deposits possible and give those deposits the maximum possible time to earn interest. That being the case, you can take advantage of the opportunities to reflect and filter in constructive memories *during* the course of a workday, or a practice session, or a workout, not simply at the end of it.

If your chosen field is a competitive sport, think about a typical practice session. It consists of a series of drills and other practice activities directed by your coaches, and those activities are how you make progress toward your goals of making (or staying on) the starting lineup, preparing for your next competition, and ultimately winning. If you're a student, think about your typical day. You have a series of class meetings, labs, and study sessions. Those activities are how you improve toward your goal of graduating with a degree. If you're one of the aforementioned "white-collar athletes," think about your typical day at the office or on the job. You have a series of meetings with coworkers or customers, separated by periods of time where you are at your desk, or on the phone, or in transit to your next meeting, and those activities are how you progress toward your goals of contributing to your organization, satisfying customer or client needs, and making a living. At the conclusion of each of these various activities taking place within a given practice or day you have the opportunity to quickly reflect, filter, and make a deposit into your mental bank account. Depending on the number of different activities that compose your day, the number of opportunities you have might be large indeed.

Let's say you're a competitive basketball player at the high school, college, or even the professional level. You arrive at practice on time, in the proper uniform, with shoes laced and with the intention of having a good practice. Coach sets the team in motion with some warm-up activities and then announces the first drill of the practice—shooting off the dribble with no defender. You get ten reps of this drill before coach blows the whistle and announces the second drill—defensive footwork on the baseline. This is the moment, as you move to that second drill but before you begin it, where you do the all-important Immediate Progress Review, or IPR. This is a minireflection on, and filtering of, what just happened during that first drill. In a similar way that you filter out the best moments or highlights of your entire day when you perform the daily E-S-P, at this moment, while moving to that second drill, you filter your ten reps of the shooting drill and focus on the best one you just had. Hold the memory of that single best shot in your mind as you jog to the baseline for the footwork drill. Make a deposit of that constructive thought and allow yourself some good feeling about it. It's just a momentary cognition, a still photo or short video in your mind's eye, a "highlight" with a very small *h*, but it's a constructive thought just the same—it's what you want more of, and the science tells us that *thinking about what you want more of is the first step in actually getting more of it.*

Now you line up for the next drill and proceed to execute, getting perhaps six or eight reps of this one. Once again the coach blows the whistle and sends you off to drill number 3—full court passing at game speed. And once again as you line up for drill number 3 you quickly reflect, filter, and lock into your memory the best rep of the drill you just finished. You may not have had any particularly great reps, but surely some were better than others and there was certainly one indeed that was either a little bit better or perhaps much better than the others. That's the highlight with a small *h*, the one you momentarily deposit into your bank account, and you run into position for that passing drill feeling just a little better about your defensive footwork.

If you were to follow this sequence of drill-filter, drill-filter, drill-filter throughout the entire practice session, you'd be effectively reviewing the progress you have made during your immediate past, hence the term Immediate Progress Review, and you'd have a dozen or so highlights with a small *h* deposited by the time coach has you doing the end-of-practice cooldown. With those dozen small highlights so deposited, once you get back to the locker room and pull out that notebook or tablet to do your daily E-S-P, you have a ton of great material to work with. Now you can easily reflect on your practice as a whole and quickly identify the day's *E*'s, *S*'s, and *P*'s. What's more, by taking the best rep of the last drill with you into the next drill, I'll bet that you had a better overall practice than you would have otherwise. Why? Because you kept yourself in a better mood by exercising constructive control over your thoughts. Sure, Coach may have chewed you out for a mistake here and there, but you were disciplined enough to *keep looking for the best in yourself* drill by drill as the practice went on.

Again, consider what will happen if you *don't* conduct this IPR after each drill. If you're anything like the hundreds of clients I've taught this drill to, you are likely to have developed the habit of taking with you not the best rep of any given drill but indeed the *worst one*. You start drill number 2 with the memory of the shot you missed, the goof you made, or the most glaring imperfection that was evident in drill number 1. Again, the science tells us that *whatever you think about most is what you are likely to get more of*, especially if that thought has a high degree of emotion to it. I'm betting that you have high expectations for yourself and that you really want to improve and succeed, but you have to be careful about what you give emotional energy to—that's what you're asking for more of.

The IPR can be utilized by practically anyone in any chosen field. The drill-filter, drill-filter, drill-filter sequence used by the basketball player can also be used by the med student, the financial analyst, and the roofing contractor. All you have to do is identify the activities that are your equivalents of the basketball player's drills. Upon finishing any class any student can momentarily fixate on the one or two points covered in the class that they are more comfortable with now than they were when the class began. When I suggest to my cadet athlete advisees at West Point that filtering their classroom experience is as helpful as filtering their sport practice, they uniformly have the "I never thought about that" reaction. But it's obvious: leaving their economics or mechanical engineering class while reminding themselves of that one principle or concept that they now understand gives them just a little more certainty about that course, and it sets them up to prepare for their next class meeting with a much better overall attitude. If, as most of them are in the habit of doing, they leave the class thinking about how hard the course is or how difficult the next exam will be, they will begin their preparation for the next day's class in a state of apprehension. Once again, there is a choice to be made—will you look for the best in yourself and your situation, even if your "best" may only be understanding the first equation the professor put on the board with the rest of the class being a total blur, or will you ignore that understanding and the benefit it represents?

What are the equivalents of "drills" in your life? What are

the discrete episodes in your workday or personal situation, each of which presents you with an opportunity for a quick reflection and the deposit of a constructive memory? In presenting this concept to a roomful of neurosurgeons it took almost no time at all to come up with a long list of possibilities: each contact with a patient, each consult with a team member, each case completed, and each filed surgical report can all be filtered and can all serve as sources of energy, optimism, and enthusiasm.

These three exercises, the Top Ten list, the Daily E-S-P, and the Immediate Progress Review, are how you can make the most of your memories, both those from long ago and those from what happened just moments ago. None of them require much in the way of time, and let's face it—they're not complicated at all. They require only your decision to look for the best in yourself day after day, and moment after moment. For all of you who grew up believing that you should remember your mistakes and imperfections as a way to achieve success, these exercises represent a big change in our thinking habits. How well are those current thinking habits working for you? Are you gaining confidence every day no matter what happens to you by thinking that way? Or to paraphrase Dr. Bob Rotella from his book *Your 15th Club*, "Is your present way of thinking consistent with the level of success you'd like to have? Does it help you find out how good you could be? And do you dare to change it?"

Let's take that dare! Start managing your memories and depositing as much "money" as possible into your mental bank account.

CHAPTER THREE

Building Your Bank Account #2: Constructive Thinking in the Present

I run a 1:56 800.

I run a 1:56 800.

I run a 1:56 800.

That was the statement Olympic hopeful Alessandra Ross made to herself each time she walked through a doorway during the nine months prior to the 2000 US Olympic Track and Field Trials. Alessandra was training to make the Olympic team in the 800 m run, a particularly brutal test of speed and guts. If you've never tried to cover 800 m just as fast as you can, imagine sprinting at your top speed for a quarter mile, then settling back to a mere seven-eighths of your top speed for the next 200 m, then putting the pedal all the way down and sprinting with everything you have left for the final 200 m. Try it sometime (assuming you are healthy enough, please!), and you will have a profound respect for the people like Alessandra Ross who ran this race at the world-class level.

With only six months to go until the Trials, Alessandra's PR (personal record) in the 800 m was 2:02.82. She kept up with the best runners in the country in training intervals and time trials, but was only ranked seventh in the nation at the time. So what business did she have stating to herself *I run a 1:56 800 m* when that was six seconds faster than she had ever run before? *She was taking the dare described in the last chapter, the dare to make the quality of her thoughts about the 800 m race consistent with the performance she desired to have, consistent with the time she ultimately desired to run.* In so doing she took the use of her "mental filter" to the next level, beyond that of ensuring constructive memories of her past, to making routine and frequent deposits *in the present*. This was her next step in winning her First Victory.

While our memories constitute a hugely important part of our mental life, and contribute, as we have seen, crucially important deposits (or withdrawals!) to our mental bank account, the thoughts we have about ourselves and the thousands of statements we make to ourselves about ourselves *in the present* are perhaps even more important. As we noted in Chapter One, fundamental to human existence is the circular process through which the thoughts we entertain about ourselves *now* (how capable we think we are, how knowledgeable or skilled we think we are) find expression in our actions and are thus confirmed and reinforced as we reflect upon them. The opinions and beliefs we hold of our talents, skills, and abilities either serve as walls that constrain us or doorways that open us up to new achievements. This chapter will show you how to extend the mental filtering process into your present moments and improve the confidence-building power of your thoughts about yourself *now*. And yes, you'll find out how Alessandra Ross did in her Olympic Trials and in some even more important trials later in her life.

When I'm explaining this cycle of thought, action, and confirmation to my cadet advisees at West Point, I typically ask them to reflect back to their very first experience in one of their mandatory first-year physical education courses, PE 117 Military Movement, or "Mil Move" as it's known in cadetspeak. The Military Movement course is described on the West Point Department of Physical Education website as "a 19-lesson course designed to expose cadets to a variety of basic movement skills. The course serves as a basis for many other athletic and military activities that cadets will encounter during their time at USMA as well as in their Army career." Cadets, on the other hand, will describe Mil Move as an exercise in frustration and exhaustion, as they are pushed through a series of graded balancing, tumbling, and climbing tasks for nineteen lessons and then on the twentieth lesson tested to see how well they can put all these skills together under the pressure of a time trial called the Indoor Obstacle Course Test, or IOCT. In this test (which they *must* pass or be recycled through the entire course again), cadets have to crawl, climb, and run through eleven obstacles at their absolute top speed to complete it in the required time of 3:30 for men and 5:29 for women. And it all takes place in West Point's oldest physical training facility, the venerable but rather dusty Hayes Gymnasium, built back in 1910, which means cadets can look forward to a unique burning sensation in the chest called "Hayes lung" as they crawl, vault, leap, balance, climb, and then finally sprint the final 350 m of the test.

"What went through your minds," I ask them, "as you lined up for that very first attendance check on the first day of Mil Move and eyeballed the mats, climbing ropes, vaulting horses, and other equipment arrayed on the floor of Hayes Gym?" I want to hear what they were telling themselves; how they were thinking about themselves with respect to this upcoming challenge. A very small minority of cadets, usually those with some gymnastics, tumbling, or climbing experience, reply, "I was thinking that this all looked like some big playground, kinda cool!," indicating a belief that Mil Move will be mostly fun; challenging perhaps, but generally enjoyable. The majority of cadets, however, especially the bigger males (e.g., recruited football players) and shorter females, respond this way: "My first thoughts about Mil Move were that I'd never done much of this stuff before, that I'm really not built for it, and so this course is really gonna suck!" Unlike the small minority who began Mil Move telling themselves that it would be exciting and challenging, this majority began their Mil Move experience telling themselves that they were in for a nasty struggle. And that difference of initial belief set into motion two variations of a universal principle in human behavior—the self-fulfilling prophecy.

The conversation with the cadets on this topic almost always proceeds this way:

ME: Based on the fact that you thought you weren't suited for all that tumbling, balancing, and climbing, how much effort and energy did you put into each class?

THEM: Some . . . Not a whole lot . . . Just enough to get me through.

ME: And what kind of grade did you finish up with?

THEM: Did okay . . . passed with a C . . .

ME: So you began the course with the belief that you weren't suited for it, and that it was going to suck.

> Not surprisingly, that belief led to a marginal effort on your part, and that effort led to a marginal grade. Guess you proved yourself right—that you're not suited for Mil Mov tasks.

And the cadet nods in agreement. I can almost hear them saying to themselves, *Yup, I was right about it the whole time.*

But then I deliver the punch line, by explaining the experience of another cadet in that same Mil Mov course, standing in the same line on the same first day, eyeballing those same mats, ropes, and vaulting horses. Unlike the first cadet who decided that he or she wasn't suited for the upcoming physical challenges of Mil Mov, this other one was thinking *This stuff is right up my alley. I'm gonna crush this.* And that belief, the thought: *I am right for this,* got his or her energy going. Thinking some variation of *I am right for this,* this other cadet had less internal resistance to attempting any of the new skills being taught and brought more persistence to getting them right, even when he or she wasn't successful with them the first few times. Again, not surprisingly, this cadet's overall greater effort, driven by that initial belief *I am right for this,* almost certainly led to a better final course grade. So this other cadet, just like the first one, proved him/herself right as well, confirming his/her initial belief.

Upon hearing all this, and seriously reflecting on it, a new understanding slowly takes hold in my cadet advisees: *perhaps the beliefs you have about yourself, expressed through what you say to yourself about yourself regarding a particular situation, are what ultimately determine what you actually experience in that situation.* Perhaps there's some truth to the old saying "people become what they think about" after all. Welcome to the self-fulfilling prophecy.

According to the *Oxford Handbook of Analytical Sociology* the term, "self-fulling prophecy," or SFP, was first coined in 1948 by American sociologist Robert Merton in reference to how "a belief or expectation, correct or incorrect, could bring about a desired or expected outcome." Merton based his observation on the earlier work of another American sociologist, William Isaac Thomas, who developed what has become known as the Thomas theorem in 1928, stating, "If men define situations as real, they are real in their consequences." Contained in both these definitions is the concept that one's ideas or beliefs about a situation, whether it's the Mil Mov class at West Point, the upcoming annual performance review at work, or a particular look on your spouse's/boyfriend's/girlfriend's face, produce real consequences. The ideas and thoughts we maintain about a situation ("I'm not built for this" vs. "I am right for this") constitute the "prophecy," a prediction about what will happen ("It's gonna suck" vs. "I'm gonna crush this"). These thoughts motivate and energize the behaviors (marginal effort vs. curiosity and persistence), which bring about the expected results, thus "fulfilling" the initial prophecy.

This fundamental fact of human life operates on many levels in nearly every activity. Students fall victim to it when they believe that they are good, for example, in math and science, but poor in English and history. Athletes fall victim to it when they are pleased with certain parts of their game (defensive play in basketball, or their forehand in tennis), but tell themselves repeatedly that other parts of their game (foul shooting or serving) are not very good. Countless individuals across the spectrum of workday activities, from computer programming to long-haul trucking and everything in between, fall victim to it whenever they allow the internal voice that says, *Oh crap, here we go again* to take over, even if it's only for a brief moment. Think for a moment of the best parts of your game, your craft, your profession, those skills or functions that you execute particularly well. Do you habitually remind yourself of how good you are in these areas and allow in some comforting feelings of competency and possibility? And conversely, do you habitually remind yourself of how much you dislike other tasks or aspects of your work, or of how ineffective you are in certain other situations? Do you realize how this is affecting you? And to repeat the question posed at the end of the last chapter, *Do you dare to change it?*

The power and pervasiveness of the self-fulfilling prophecy has been known throughout human history. The Greek myths of Oedipus and Pygmalion teach us how our underlying initial beliefs about ourselves and others can have both tragic (for Oedipus) and triumphant (for Pygmalion) consequences. In the second century BC the Roman emperor Marcus Aurelius recorded his own understanding of the self-fulfilling prophecy in a series of reflections and essays on self-improvement, later published under the title *Meditations*. In it he observed that "Our life is what our thoughts make it" and "The happiness of your life depends upon the quality of your thoughts." The Book of Proverbs (chapter 23, verse 7) in the King James Bible reminds us that "As a man thinketh in his heart, so is he." Shakespeare crafted his famous Macbeth character around a prophecy that a king unwittingly fulfilled, leading to his tragic demise. In the nineteenth century, American transcendentalist Ralph Waldo Emerson penned the phrase "A man is what he thinks about all day long" as he lectured for the abolition of slavery, Native American rights, and the general improvement of mankind. More recently, New Age writers like Marianne Williamson and Wayne Dyer have encouraged their readers to carefully examine the stories they tell themselves and their own self-constructed narratives.

New or old, classical or modern, all these expressions are variations on one theme: we are telling ourselves stories about ourselves practically every waking minute, establishing various prophecies that we then act, almost automatically and unconsciously, to fulfill. These stories range from reminders of what we need to do next or what we should have done before, what we are good at, bad at, right about, wrong about, and on and on and on. Each one of these stories, every statement we make to ourselves, enters into the mental bank account of our confidence and either drives it upward or drags it down. Winning the First Victory, then, involves first becoming aware of the stories and statements you make to yourself about yourself, the dominant narrative that you use to define, reinforce, and motivate yourself, and, second, exercising the discipline to ensure that the stories and statements you do make to yourself meet the criteria we established in the last chapter for passing through the mental filter—creating energy, optimism, and enthusiasm.

More than thirty years of careful psychological research has shown that when people affirm their value, when they incorporate into their personal story lines particular constructive thoughts about themselves *in the present*, they maintain, in the words of Stanford psychologist Geoffrey Cohen and UC Santa Barbara psychologist David Sherman, "an overarching narrative of the self's adequacy." Their research has found positive effects of self-affirmation on behavioral changes across a wide spectrum of activities, including smoking cessation, academic performance, interpersonal relationships, and weight loss. Individuals who establish this greater sense of personal adequacy through self-affirmation techniques persevere with learning new skills and more successfully cope with setbacks. From this research, Sherman concludes, "Self-affirmation can thus lead to self-improvement in terms of less defensiveness and stress and more positive behavioral change and better performance."

Harvard University psychologists Alia Crum and Ellen Langer came to a similar conclusion from their research into the effects of a shift in mindset on the health of hotel workers. Simply by changing their thinking from "I don't get much regular exercise" to "I'm getting regular exercise cleaning fifteen rooms every day," forty-four hotel workers lost an average of two pounds and decreased their systolic blood pressure by ten points in a month's time. A matched group of workers, doing the same jobs in the same hotels, but who were not taught to think about their work as exercise, experienced significantly fewer physical changes over the same time period. Both groups of workers reported getting no additional exercise outside of their work, and neither group increased the volume of their work activities or the speed at which they completed their appointed work tasks. Crum and Langer conclude, "It is clear that our health is significantly affected by our mindsets."

Pause for a moment right now and consider what stories and statements that voice (or chorus of voices) from the back of your mind has been telling you throughout the course of the day so far. Did that voice whisper, *You better not mess this up* as you approached an important encounter? Did that voice whine, *Why are the coaches always in my face?* as you changed after a hard practice or training session? Did that voice scream, *You idiot! My grandmother could've hit a better shot!* when the tennis ball zoomed out of the court? Or did that voice steadily and consistently affirm (that is, *make firm in your mind*) a desired feeling, quality, or outcome that you wished to experience at that moment? Did it provide encouragement as you approached that important encounter with the statement *I meet each new situation with determination and understanding*? Did it help maintain a constructive perspective after that hard practice with the statement *I get better from each new coaching point*? Did it keep you focused after that missed tennis shot with the statement *Once a point is played I move right on to the next one*? It is *affirmation statements* such as these, statements we make to ourselves about *the reality we wish to experience, but phrased as if that reality is happening right now in the present*, that we can use throughout the day to make frequent deposits into our mental bank account and turn the universal self-fulfilling prophecy from a thief and enemy into an ally and partner.

Making the Self-Fulfilling Prophecy Work for You: Writing Out Your Mental Deposit Slips

Now that you've learned that the statements and stories you tell yourself about yourself are influencing the course of your life right now, that they influence the energy and effort you put into various tasks, functions, and behaviors, that they influence how you respond to setbacks and even change your biology, it's time to get busy and start using them to win your First Victory. Time to start talking constructively to yourself about how you want to perform and how you want to be, not in some yet to be determined future, but right now in the present.

Start by thinking about a skill, or a quality, or a characteristic that you have right now that you are relatively pleased with. Hopefully you can come up with a few . . .

Let's say you're a competitive hockey player. You might come up with *I have a fast and accurate shot* or *I am a solid defensive player.*

Let's say you're a recreational golfer. You might come up with *I read greens pretty well* or *My midrange irons are pretty consistent.*

Let's say you're in leadership position as an executive. You might come up with *I handle disagreements thoughtfully* or *I communicate our team vision well.*

Congratulations, you have just composed your first affirmations.

Now notice the structure of each of the above statements. You see that they are all personal—meaning they are built around "I," the first-person singular pronoun, or "me," or "mine," references to the singular person that is you. This is important. Building confidence is about building up *your personal* mental bank account, about *your total of thoughts about yourself,* which means that a general, relatively impersonal statement like *Handling disagreements thoughtfully is good* won't have much impact on your confidence one way or another. That statement, while perfectly true, isn't focused on *you*, the person whose confidence you are building up. For an affirmation to work, for it to actually build up that mental bank account, it has to be *personal*, as in *I handle each disagreement thoughtfully* (Note: the first-person plural pronoun "we" can be used when developing affirmation for a team with a shared goal or target, as in "We bring a unified intensity each time the ball is put in play" or "Our collective experience allows us to handle any customer issue").

You will also see that each of the above statements is made in the *present tense*—meaning that they are expressions of what is happening *now* rather than in some hoped-for future. This is also important. The balance of your mental bank account is what's in it right now, not what you hope to have in it someday. Using the future tense and telling yourself *I will be a better listener in team meetings* won't do much for the sense of certainty you are trying to build and maintain, as it tends to remind you of what you *are not now.* I've noticed over the years that most performers whose self-talk includes a lot of future-oriented language like *I will develop this skill . . .* and *My execution will improve . . .* are habitually putting real change off into the future, essentially delaying it over and over again. So put some urgency into the affirmation process by keeping everything in the *present tense.* Right now, this moment, is really the only thing any of have. Get busy using it!

Finally, you will see that each of the above statements is phrased *positively*—meaning they state *what you want more of* rather than emphasize what you want less of. An effective affirmation statement affirms, that is, asserts or confirms, something that is desirable and valuable, rather than minimizes, negates, or denies something you are trying to avoid.

This is a crucial difference that many people fail to understand and take advantage of. The tennis player's statement *I never miss my second serve* might not initially sound any different from *My second serve hits just inside the service line,* but it produces a significantly different impact at the neurological level. Apparently the part of the brain that controls the execution of that tennis serve doesn't do a good job of differentiating between "never missing the second serve" and indeed "missing" that serve—it only recognizes the active verb "missing," and activates the neural pathways associated with memories of missed serves. Each and every repetition of the thought *I never miss my second serve* essentially asks the brain to retrieve those memories again and again, activating the associated neural pathways each time. Unfortunately, with each activation those pathways run a little faster, smoother, and stronger. All this results in a startling neurological fact—thinking about what you don't want to happen only reinforces your brain's familiarity with it, making it only more likely that it will actually happen. So while on the surface the statement *I never miss my second serve* seems like textbook "positive thinking," it's actually reinforcing something undesirable and carving the neurology of that undesirable behavior deeper and deeper into your nervous system. The more constructive alternative is obvious—carve into your brain the neurology associated with what you want more of—placing that second serve right inside the service line, reaching consensus with your work team, accurately reading the body language of your customers—by talking to yourself in the present tense about what you want more of, phrased as if you already have it or as if it already exists.

The implications of this are clear for anyone pursuing his or her First Victory—by repeating a story about yourself to yourself that is personal and positively phrased in the present tense you make deposits into your mental bank account and build your sense of certainty. Affirming your skills, your capabilities, and your positive characteristics changes how you see yourself in relation to your chosen profession, sport, or craft and initiates a constructive self-fulfilling prophecy. And that's only the start.

As powerful as affirmation statements may be for defining and reinforcing what you have now, their real power lies in what they can do for you about the skills, capabilities, and characteristics you do not yet possess, and the outcomes you are currently pursuing but haven't yet achieved. Alessandra Ross, the Olympic hopeful we met at the opening of this chapter, had a personal best of 2:02 in the 800 m run, a respectable time to be sure, but not the one she wanted. Instead of fixating on that present best time, she invested in the simple affirmation "I run a 1:56 800 m," repeating it to herself many times a day. While some might say this was merely wishful thinking on her part, the science we've reviewed so far suggests that she was on the right track (no pun intended). By affirming that desired time, by in essence saying yes to it (to "affirm" is to say yes), Alessandra Ross was giving herself a better chance of finding out just how fast she could run than she would have by being strictly realistic and only believing in what she had already accomplished.

Take a moment and consider your own personal equivalent to Ms. Ross's desired outcome—running the 800 m in 1:56. What's something that you would love to achieve in your professional life or in your personal life? Are you affirming it? Are you saying yes to it now in the present moment? Going one step further, are you affirming (saying yes to) any improvements or changes in your skills and abilities that would help you get to that desired outcome? This is the next step in your confidence-building regimen—a steady diet of effective affirmations that will make multiple deposits into the mental bank account every day.

How might this look? Let's use our Olympic hopeful Alessandra Ross as a model. Her overall affirmation "I run a 1:56 800 m" was supported by a series of affirmations regarding both her training and her actual racing. Here's a sample:

- Every day in practice I am tuned in to my form and stride.

- I complete each interval at the time coach sets out.
- I'm so good that I can even be sick for a couple weeks and still turn in a great race time.
- My mindset during the warm-ups and the moments before the race is total excitement.
- I am more excited than ever about my next opportunity to go 1:56 in the next big race.

Here's a sample of the affirmations used by another athlete, wrestler Phillip Simpson, in pursuit of a national collegiate championship:

- I drill with a purpose every day.
- I treat every practice as a precious opportunity to improve.
- I get to bed by 11:30 six nights a week to allow recovery.
- I spend fifteen minutes building explosive strength every other day after the team's session.
- I am in the best shape of my life and I love it.
- I am on that winner's platform with my hand raised to the ceiling.

And here's a sample of the affirmations used by a white-collar athlete to build his confidence in moving up to the publisher's level in the magazine industry:

- I get out to a minimum of ten sales calls a week to listen and learn about the issues that affect the sales force.
- I create the culture where every sales rep proclaims the unique value of the publication.

- I lead with calm determination no matter the situation.
- I am the leader that gets the team to believe in the possibilities for our magazine.

Now it's your turn. Get out something to write with and compose three statements that affirm some of your best qualities and skills, or some effective action or actions that you take. Use the examples above as guides and write your affirmations with the following five rules:

1. In the *first person*—"I participate in the daily functions where *I* can be visible and influential" vs. "*Good leaders* are visible and influential."
2. In the *present tense*—"I *enjoy* competing in close games" vs. "I am *going to* get better at playing close games."
3. With *positive* language—"I *organize* my activities and time *effectively*" vs. "I *no longer waste* my time."
4. With *precise* language—"I run *four miles three times a week at a seven-minute-per-mile* pace" vs. "I run *regularly*."
5. With *powerful* language—"I *explode* from any position *like a bullet through a gun*" vs. "I *come out hard* from any position."

Now take the next step toward your First Victory by writing three more affirmation statements. Write one about *a quality or skill that you do not yet have but wish to develop* ("My confidence stays high when I sense that my team isn't understanding me"). Write a second one about *an action that you do not presently take but that you know would be helpful to you* ("I get out to a minimum of ten sales calls a week"). Write a third one about *an outcome you wish to achieve but have not experienced* ("I am on that winner's platform with my hand raised to the ceiling"). Write these statements according to the same five rules—first person, present tense, positive, precise, and powerful language. Phrase each one as if you already have that quality, already take that action, and have already achieved that desired outcome.

This exercise may take you way out of your personal comfort zone. You may feel decidedly uncomfortable stating, *I arrive at the gym three times a week at 8 a.m. ready to push myself hard* when the best you've done over the last six months is twice a week. Perhaps some internal voice is telling you that you're just lying to yourself by thinking about something you've never done in present tense and positive terms. Let's say your business is still in the idea phase and while you'd certainly love to have a net positive monthly cash flow, you don't think it's at all realistic to affirm *I achieve a net positive monthly cash flow* by a certain date. My clients routinely raise this issue with me, asking, "Shouldn't I be more realistic? Aren't I just fooling myself when I do this?"

When I get this question I pull out an old grainy video clip that I found way back in 1992. It shows the 1986, 1989, and 1990 Tour de France champion, Greg LeMond, pumping his bicycle up a steep hill in a scene shot for this video. As the video plays for just over a minute, showing LeMond power his way past a pack of other cyclists, the sound track plays LeMond's accompanying internal dialogue, and that voice emits a constant stream of affirmations, "Now the hill is easy . . . My legs are strong . . . my back is strong . . . there's no effort at all . . . it's as easy as breathing." After showing the video clip, I ask my audience if anyone thinks Greg LeMond actually feels that good while pushing himself up a steep hill at a fast pace. The answer is always "Of course not!" And the audience is right—LeMond certainly doesn't feel perfectly strong or perfectly effortless at that moment. However, and this is the key point, instead of fixating on whatever discomfort or effort he may actually be experiencing, LeMond is affirming (saying yes to), the *reality that he wants to have*, the sensations and comfort level that he ideally wants to experience at that moment. As that stream of personal, present tense, positive, precise, and powerful statements continues, LeMond wins his First Victory over fatigue and self-doubt. He establishes a constructive self-fulfilling prophecy for that moment of effort on that hill in that race, and by doing so, he encourages his heart, lungs, and muscles to continue to function at optimal levels (recall the hotel workers who lost weight and reduced their blood pressure by believing that they were getting regular exercise). Try out your own version of LeMond's affirmation stream during your next trip to the gym or on your next training run. As you puff away on the stationary bike, elliptical trainer, or treadmill, or as you pound the pavement, talk to yourself about how you want to feel as if you are actually feeling it: *My breath is steady and full . . . my strides are smooth and powerful . . . I love getting my heart rate up and my blood flowing . . .* I have never known anyone to come back from such an experience and report that they didn't feel better throughout their workout and didn't perform a little bit better than expected, simply from aligning their thoughts in the moment to the performance they wished to have. First Victory won.

This understanding shifts the question from "Am I being realistic? Am I lying to myself?" to "Am I doing what I can to help myself (and my team) in this moment right now?" Are you actually being realistic or are you simply justifying some negative thinking? The basketball player who claims, "I'm just being realistic about myself" when he hesitates to state to affirm *I am 100% dependable at the foul line* is simply maintaining a negative self-image based on his memories of missed shots and unwittingly engaging a negative self-fulfilling prophecy. If this same player were to look objectively back through his entire personal history playing the game he would certainly find a memory or two of making foul shots. The shots he made are just as "real" as the ones he missed and can serve every bit as much as the basis for his self-image and a more constructive self-fulfilling prophecy as the misses. He's no different from the West Point cadet who dreads the crawling, climbing, and balancing tasks of the required Military Movement course. Both of them may be absolutely right about themselves at the moment. They may both indeed have failed or performed poorly at the foul line or in some gymnastic task at some time or times in the past. That's truth, no denying it. But does that mean they can't do it now? Are they being "realistic" or inadvertently *choosing to be negative*? Without getting too deep into existential philosophy and the questions about what *reality* is or is not, we all experience the world and our lives through a deeply personal lens that shapes our unique perceptions and determines what is real for us at any moment. Something as simple as tomorrow's sunrise can be perceived as the start of yet another threatening episode or as the beginning of a long period of opportunity. Either one can be *real*, but which one do you want? Which one serves you best, either in the short term of today or in the long term of a thousand tomorrows?

Getting the Self-Fulfilling Prophecy to Work for You: Making Frequent Deposits

Option #1: The Notebook Nightcap

American speed skater Dan Jansen was about to skate his last event in his fourth Olympic Games. After disappointing results in Sarajevo in 1984, the tragic death of his beloved younger sister as he was competing in Calgary in 1988 leading to falls and unfinished races, and more disappointing results in Albertville in 1992 (fourth in the 500 m and twenty-sixth in the 1,000 m), Dan Jansen was something of a question mark. How come this guy who has skated world-record times and won world championships can't seem to get it done in the Olympics? But in that last Olympic race, the 1,000 m in Lillehammer in 1994, Dan Jansen answered with an unexpected gold medal performance and a world record time of 1:12:43. What made the difference? At least one contributing factor to his success was the work Jansen did with sport psychologist Jim Loehr, "to get my mind to actually like the 1,000 m race." Jansen recalls that "I almost feared the 1,000 before; I knew after doing it so many times that I would tire in the last lap, and I almost came to expect that I would get tired."

Knowing that the expectation of getting tired was a potentially disastrous self-fulfilling prophecy, Jansen and Loehr got busy. "So we did all these crazy things," Jansen recalls, "like writing down every day 'I love the 1,000.'" In fact, Loehr had Jansen write that "I love the 1,000" affirmation in a notebook a dozen times each night before retiring for the two years leading up to that final Olympic opportunity in 1994, effectively setting into motion a positive self-fulfilling prophecy that would promote energy and enthusiasm instead of worry about tiring in the final lap. (I know Jim Loehr, and he insists that his clients commit to this kind of serious practice.)

I recommend this "Notebook Nightcap" for making multiple deposits in the mental bank account to all my clients. Dan Jansen's dedication to his Notebook Nightcap resulted in 8,670 deposits over two years. Finish out your day by writing each of the three affirmations you have composed in a notebook or journal *at least* three times each. Anyone who is serious about winning their First Victory can devote five minutes before retiring to this practice. As you write each of your affirmations, let them create a strong internal feeling. If you are affirming a skill or action ("My quick feet match any opponent"), feel yourself doing it. If you are affirming an outcome or achievement ("I am the 2020 Sales Professional of the Year"), feel the sense of accomplishment that outcome would bring. Making the last thoughts of your day personal, positive, and powerful gives the subconscious parts of your mind useful material to process while you are sleeping, without interference from your conscious mind. You might even find that ending your day with these good feelings promotes a more peaceful sleep.

Option #2: The Open Doorway

How many times a day do you walk through a doorway? When I ask cadets at West Point this, their eyes swim and they blurt out answers like "Lots! . . . Too many to count! . . . I have no idea but I'm sure it's huge!" What if every time you walked through any doorway, you used it as an opportunity to make a deposit into your mental bank account by repeating your three key affirmations to yourself? How many deposits into your mental bank account would you be making if you used every doorway you passed through as a trigger to reaffirm that desired quality, that desired action, or that desired outcome? Welcome to the "Open Doorway" exercise.

Alessandra Ross chose this exercise and used it consistently in the nine months before her Olympic Track and Field Trials in 2000. While she never kept count of the number of doorways and hence the number of times she repeated her *I run a 1:56 800 m* affirmation, some quick math gives us an idea. If you walk through an average of fifty doors a day (go ahead, count them; this is a realistic number), for a period of nine months that comes to 13,500 deposits (50 a day X 270 days). She chose this method on my recommendation—why limit affirming a desired quality or outcome to one specific time of day? Although many therapists and life coaches prescribe repeating affirmations in the relatively relaxed moments right before bed so that the mind has quality material to process during sleep (and there's certainly no reason not to do this, as the Notebook Nightcap exercise suggests), I know of no scientific data that supports limiting the affirmation process to any particular time of day. Why not max out the number of deposits you can make in a day? Think about how many times a day you worry about a present problem or an upcoming performance. Why not turn the tables on that?

While I never asked Ross how often she worried about not achieving her Olympic dream, she admitted that she had plenty of self-doubt, and she knew it was affecting her times on the track. In that regard she was entirely normal. But unlike the many athletes who deny their self-doubt and refuse to acknowledge that they've engaged an unproductive self-fulfilling prophecy, Ross decided to do something about her situation and began using the Open Doorway to build her confidence. Getting started with this as a regular practice was a challenge. Like so many athletes who struggle with routine, everyday self-doubt, Ross was initially uncomfortable with affirming her desired time of 1:56. *Who am I to think that I can do that?* she thought at first. "Who are you NOT to think that?" I told her during one of our early meetings, showing her the passage "Our Deepest Fear" from writer Marianne Williamson. "Three athletes are going to qualify to run the women's 800 for the U.S. Olympic team," I continued. "One of them might as well be you." So Ross complemented her rigorous physical training by reexamining her attitude and her thinking habits. She soon realized that her overserious perfectionism and constant comparisons to other racers were hampering her development, and that it was, in fact, okay to think about what she wanted. She opened up her mental bank account and started building it up.

Within a month of repeating her affirmation every time she walked through a doorway, Ross found herself becoming more and more comfortable with the self-image she was creating. Her initial leap of faith into the idea of running 1:56 morphed into a stronger and stronger sense of self, to the point where she felt herself *becoming* a 1:56 runner, even though she was not yet running that time in a race. (See Appendix I for the personal script I created for Ross to listen to during these months.) The self-doubt that she had previously carried into races throughout her college career gradually faded away into irrelevancy. Her new sense of certainty, her hard-won First Victory, paid handsome dividends in the 2000 Olympic Trials, where Ross set two consecutive personal best times, the only contestant in her event to do so, and earned the alternate spot on the 2000 Olympic team. Her final time of 2:01, while not the 1:56 she had been affirming, was her best ever, and she achieved that PR (personal record) despite two significant and unexpected disadvantages: having to start the race in the innermost, least advantageous lane, where she had to run through the tightest turns instead of striding out fully, and having to hurdle over a boom microphone that had been mistakenly extended out into her lane. While she never did run that 1:56, there is no doubt that Ross won her First Victory.

And that was just the start. Ross went on to graduate from Georgetown University medical school through a US Army Health Professions Scholarship and then complete her residency in orthopedic surgery, one of only a tiny handful of women to do so (as of 2018 only 6 percent of board-certified orthopedic surgeons were female). Succeeding in this demanding field required the same level of confidence that helped her make the Olympic team, so she continued talking to herself about how she wanted to be and what she wanted to accomplish as if it were all happening right now. To help herself deal with the stress and lack of sleep she affirmed, *I live happily in the awareness that I control my thoughts and therefore my destiny . . . I remain calm when confronted with a difficult situation . . . I catch any self-criticism and instantly throw it away.* To help herself memorize anatomy she affirmed, *I quickly name muscle origin, insertion, innervation, and function . . . My surgical exposure and intra-operative anatomy is solid . . . I fluidly present the knowledge I've obtained.* And to help her rise above the hostile, male-dominated work environments she encountered, she affirmed, *I'm fully prepared for my cases with every diagnosis and treatment option . . . I'm comfortable with what I do, who I am, and why I chose this specialty.*

Following the completion of her residency and internship, Ross served for six years as an orthopedic surgeon for the US Army before retiring in 2014. She served at Fort Jackson, South Carolina, at the 121st Combat Support Hospital in Seoul, South Korea, and in May 2011 was deployed to a NATO hospital in Afghanistan for six months. There, in the midst of horrific carnage, spending her days repairing the limbs and lives of wounded soldiers, she called upon the same mental resources that had sustained her throughout her athletic career and so far in her medical career. She continued to affirm her value, worth, and dignity. She had known, even as she left her husband and two young children for Afghanistan, that this would be the performance she had been preparing for all her life, and that it would be the greatest test of the personal courage and confidence that she had built and protected for over a decade. Perhaps not surprisingly, when her deployment was completed, she found that she had grown and benefited from the adversity she experienced rather than being diminished by it. As she told me years later, "I came back knowing the parts of myself that I could rely on, and all the work we did for track, medical school, and residency laid the foundation for that."

While much has been deservedly written about the devastating effects of post-traumatic stress in the aftermath of the American military's presence in the Middle East, and while thousands of American servicemen and -women suffer from it every day, less has been written about *post-traumatic growth*, the process of finding a new appreciation for life and a greater sense of purpose following adversity. Developed by psychologists Richard Tedeschi and Lawrence Calhoun in the mid-1990s, post-traumatic growth (PTG) theory holds that people who endure psychological struggle can often see positive growth afterward. To quote Tedeschi from a 2016 *American Psychological Association Monitor* article, "People develop new understandings of themselves, the world they live in, how to relate to other people, the kind of future they might have and a better understanding of how to live life."

Alessandra Ross came back from her Afghanistan deployment with precisely that new understanding. Ever since returning home she has served patients, created empowering networks for other female surgeons, and raised her children to believe that they are in charge of their thoughts. Completely comfortable with the idea of telling yourself you have achieved something even if you haven't, Alessandra Ross continues to build her confidence with affirmations each time she walks through a doorway. If you are close enough to her when she does so, you might hear her whisper, *"I am Radiant!"*

Just as each doorway you encounter as you go through the day takes you from one physical space to another, whether it's to another room or into or out of a building, that doorway also takes you into a new moment, into a new personal space, into a new *now*. What will you affirm, what will you say yes to in that new *now*? What self-fulfilling prophecy will you take with you into that new *now*? Pass through each doorway affirming that desired quality, that desired action, and that desired outcome.

Option #3: The Macro Affirmation Script and Audio

Cadet First Class Gunnar Miller was in my office late in October 2016, looking back on his performance in the army men's lacrosse fall training period, and he wasn't happy. A standout midfielder from Upstate New York, where lacrosse matters more than any other sport, Gunnar was a high school all-American, and Offensive Player of the Year in the very competitive Section V. But to listen to him on that day you would have thought he was the worse player ever to suit up for West Point. It was immediately clear how ineffective Gunnar's thinking was ("My lacrosse IQ is so high that when I make a mistake I get really upset"), how poorly his mental filter was working ("I can recall specific mistakes I made five weeks ago with total detail"), and that his mental bank account was down to nothing ("I didn't produce the way I should . . . I didn't feel confident when dodging to the goal with the ball in my stick").

So we got busy. I explained to Gunnar the connection between his conscious thoughts and his unconscious execution on the lacrosse field. Together we investigated the way his thoughts and memories had apparently affected his skills throughout the fall training period and were still affecting his skills right now. Gunnar readily caught on to the bank account/confidence metaphor and admitted that he was only giving himself 60 percent constructive input, a level of success that would earn him a D grade in any West Point course. I immediately had him recall three personal highlights from the team's fall scrimmage. Not surprisingly, he came up with them in a matter of minutes, and his face lit up as we wrote them out on my office whiteboard. Memories like these, I explained, make up a quality mental diet, and I directed Gunnar to come back to our next meeting with the memories of (1) three instances of quality dodging to the goal, (2) three quality shots on goal, (3) three ground balls successfully picked up, (4) three instances of quality defensive play, and

(5) three instances of quality off-ball movement. Performers who are committed to their success, I have found, don't mind doing a little homework.

Gunnar returned to my office five days later, having spent fifteen minutes completing the following list:

Dodging:

- Move vs. Navy in Patriot League semifinal game for a goal
- Move vs. Navy freshman year to take a lefty bounce shot
- Roll dodge on Loyola's all-American defensive midfielder for a goal

Shooting:

- Rip vs. Colgate in the Patriot League Championship
- Down the right-hand alley vs. Loyola in Patriot League quarterfinals
- Freshman year high to high stick side vs. Holy Cross

Ground Balls:

- Tough GB vs. Michigan leading to a goal
- Tough GB vs. Holy Cross leading to a goal
- Defensive stand vs. Lehigh with two GBs leading to a clear

Defensive Play:

- Vs. Lehigh during a long possession, then cleared the ball
- Matched up vs. our best dodger in practice denying him a single shot
- Stripped the ball from our starting attackman in practice

Off-Ball Movement:

- Crease pick for Nate, then rolled off to catch the ball and throw a behind-the-back feed to Cole
- Re-picked for Cole then rolled off and scored on a quick release
- Seal picked for Dave to a score

These memories opened up Gunnar's mental bank account, and he was off and running. We spent the next hour discussing how to continue feeding his confidence through the daily E-S-P exercise and extend his mental management into his thoughts about himself in the present. Gunnar immediately got busy with his own version of the Open Doorway exercise, making use of each time he entered and exited West Point classrooms throughout his day. Then, during our next meeting, I brought out the affirmation equivalent of heavy artillery: the construction of a comprehensive and personalized affirmation script that would be recorded into an MP3 audio file. I had been making these custom audio products for clients for over twenty-five years, and Gunnar Miller jumped at the opportunity. This "Macro Affirmation Script and Audio" would provide Gunnar with an uninterrupted ten-minute narration of affirmations, a bombardment of statements phrased in the first person and present tense, written with positive, precise, and powerful language, complete with inspiring music in the background. Here is the first section of that Macro Affirmation Script:

> I am a deadly dodger from anywhere on the field . . .
> I love dodging in the biggest games and against the

> best opponents . . . I create opportunities every time I dodge . . . I get my hands free for a dead-on feed or a nasty shot over and over again . . . If I happen to get stood up on a dodge, I forget about it and get ready for the next one . . . I see the defenseman, I see the goal, and I let it happen . . . Each stride is effortless . . . My feet feel light on the ground . . . I feel so confident in my ability, there's no one who can guard me . . . I am a deadly dodger from anywhere on the field . . .

Gunnar's complete script consisted of similarly detailed paragraphs on both the technical aspects of his play (dodging, shooting, ground balls, off-ball movement, and defense) and some mental aspects of his play (handling inevitable setbacks and maintaining a sense of himself as a winner). It concluded with the paragraph:

> This is how I think from here on in . . . This is how I'll take my game to a new level. . . . I am proud to be part of army lacrosse and I accept the responsibility that honor brings . . . It's up to me to make the best of this opportunity . . . I'm gonna do what I have never done before, so that I can be better than I've ever been before . . . and when it's over, my game will be at a whole new level . . . Let's go, this is my time to shine!

I let Gunnar "test drive" the completed audio version of his Macro Affirmation Script in my office, reclining back in a comfortable chair with his eyes closed. Ten minutes later, as the last strains of the final music selection faded out, he opened his eyes and grinned a grin I hadn't seen in months. When I asked him how he felt at that moment, his response was "Really fired up!" Having just saturated his mind with statements about great play, his emotional state was understandably one of eagerness, excitement, and, of course, confidence. "I wish I could play right now!" he declared. First Victory won.

Gunnar put his Macro Affirmation Script and Audio to work throughout the winter of 2016–17 and all through his 2017 lacrosse season. Listening to the audio became a daily ritual on the bus ride up to the stadium for practice, and part of his pregame mental preparation. That 2017 season was his best ever. Chosen to be captain by his teammates, he started in all sixteen games, tallied game-winning goals twice, played a key role in the team's upset wins over Syracuse and Notre Dame, and was selected to the Patriot League All-Tournament Team. In our final year-end closeout meeting, when I asked him what suggestions he had for me to improve my work, Gunnar's only comment was "Have Coach mandate every guy on the team to have his own affirmation script and audio!"

As of this writing First Lieutenant Gunnar Miller is an executive officer in a Basic Training Brigade at Fort Jackson, South Carolina. All army jobs, as any soldier or officer will tell you, come with some problems and complications. Miller's is no different, but just as he rose above the difficulties of that fall training period in 2016 by getting control of his thoughts, he's doing the same today. Each time he walks through a doorway he repeats, *I have a new opportunity to represent the name on the front of my chest . . . I am in the best shape of my life . . . and I get to see the woman I love every day.* And his Macro Affirmation Script audio is still on his cell phone. (Readers wishing to get their own Macro Affirmation Script and Audio made can contact me at NateZinsser.com for details.)

Just as we concluded the last chapter with the question "Is your present way of thinking consistent with the level of success you'd like to have?" we conclude this chapter with another one: "Who do you think you are?" What are the ongoing stories you are telling yourself about yourself *right now*? Again, are those stories consistent with the level of success and satisfaction you wish to have? Do you love your personal equivalent of Dan Jansen's 1,000 m race, whatever that contest, test, or performance may be? Are you fully prepared to deliver each of your personal workday equivalents of a surgical resident's diagnosis and treatment options? Whatever you choose to believe about yourself will find its way into your actions and eventually into your outcomes. The exercises presented here, the Notebook Nightcap, the Open Doorway, and the Macro Affirmation Script and Audio all provide tools for the telling of the constructive stories that will harness the power of this universal self-fulfilling prophecy and make it work for you. Do you want to be *Radiant*? Affirm it first, and then don't be surprised at finding yourself becoming it.

CHAPTER FOUR

Building Your Bank Account #3: Envisioning Your Ideal Future

Colonel Kevin Capra, a protégé of mine from the West Point Class of 1995, commanded the Third Armor Brigade Combat Team of the Army's First Cavalry Division at Fort Hood, Texas, from July 2018 until June 2020. The brigade consists of thirty-seven companies, a total of 4,300 soldiers, (a company is a military unit of 80 to 120 soldiers subdivided into smaller units called platoons). Each company is led by a captain who's in charge of the training and preparation of that company for three years, at which point he or she will rotate out to a new assignment and a new company commander will take over. During Colonel Capra's assignment to the brigade, he would take each new company commander through an exercise of the imagination.

He'd ask, "What kind of training events and environments do you want your soldiers in?" And every new commander answered with the word *realistic*, because each one knew that training for combat has to be as close to the real thing as possible. Contrary to what we see in movies and TV shows, soldiers do not heroically rise up to new levels of performance in combat. Instead, they sink back to the level of their training. Then Colonel Capra got to work, getting down to what that word *realistic* means.

"Can you picture that kind of realistic training event?"

"Can you hear it—the weapons firing, the communications happening, the explosions going off?"

"Can you feel it—the radio in your hand, your movements on the ground or in your vehicle, the sweat streaming down your back?"

"Can you smell it—the cordite, the sand, the wind?"

"Can you taste it—the grit and the blood in your mouth?"

These questions, and the conversations that followed, set the conditions, the timelines, and the resource decisions necessary to train that new commander's company into the best in the world. For Colonel Kevin Capra, *imagine* is the most powerful word in the English language.

Former American long jump specialist Mike Powell would stand in his living room, waiting for the room to become cool and dark. That way, as a 1994 *Sports Illustrated* article reported, "he can see the dream better." Powell strode through the room, turned left through the dining room, and as he stepped into the foyer he imagined jumping and breaking the world record set by Bob Beamon in the 1968 Mexico City Olympics, the longest-standing record in track and field. Powell's flight of fancy always ended with him throwing his hands up in celebration as he heard the cheers of the crowd and experienced the elation he knew that moment will bring. "I could feel it in my head," he recalled. "I've envisioned that jump a hundred times."

Second Lieutenant Paul Tocci, another West Point protégé

(class of 2016), would sit down to repeat a nightly ritual. Having finished his various tasks and trainings for the day at his Basic Officer Leadership Course (where all freshly minted lieutenants go after earning their commission), Paul settled down into the most comfortable chair he could find, put on his headphones, and opened up a sound file on his phone. That file began with a voice providing some simple instructions to relax and focus: "Assume a comfortable position . . . direct your attention to your breath . . . feel the air entering and leaving your body . . ." After four minutes of this guidance, the voice shifted from guiding Paul's body to guiding Paul's imagination, taking him on a trip into the entrepreneurial future of his dreams:

This is my chance, the opportunity to get to a place that few ever dream about . . . It's my time to jump into an elite tier of successful people . . . I'm on track to become the most successful entrepreneur to ever graduate from West Point . . . I will accomplish this by onboarding 80 percent of the Ft. Leonard Wood population onto the Trade U platform by October 15, and by achieving a net positive monthly cash flow by December 1, . . . in a year I'll generate the perfect statistical profile to attract new opportunities within the Army and investors for my technology . . .

What were these guys doing? Idle daydreaming? Was Colonel Capra playing some grown-up version of "let's pretend" with his new company commanders? Hardly—Capra, Powell, and Tocci were all exercising their mental filters in yet another powerful way: making deposits into their mental back accounts by using a particular thinking process that involved all their senses—vision, hearing, smell, taste, touch, position, and movement. I call this process *envisioning*, the *deliberate production of an emotionally powerful, multisensory imaginative experience of a de-*

sired future event or events. Having covered (1) the management of memories from your past, both long-term and recent, and (2) the control over the stories you tell yourself about yourself in the present, we turn now to controlling the visions, the pictures, and the feelings you create and maintain about your future. That's right, we're talking about your imagination, that uniquely human mental function through which you "see" into your future. You employ that imagination of yours every day when you plan out your class, work, or training schedule, when you contemplate your long-term future, and also (unfortunately) when you worry about all the things that could go wrong in your world. By the time you finish this chapter you will know how to use your imagination selectively and constructively to help you build more confidence and become more "mentally prepared" to perform at your peak levels. Just as we have seen the value of taking control of our memories of the past and taking control of the stories we tell ourselves about who we are in the present, there is also great value in taking control of our thoughts about the future through the mental skill of *envisioning*.

You've probably heard about or read about athletes using "visualization" to prepare for an upcoming competition; the gymnast running through his floor exercise routine in his mind before executing it, the soccer player imagining the thrill she will experience when she scores the winning goal on a penalty kick, or the tennis player recalling the exact movements that produced a winning shot. Athletes and other performers—actors, musicians, surgeons, and salesmen—have long practiced their own varieties of "visualization" under various names: mental rehearsal, motor imagery, creative visualization. Even the army has long practiced what it calls "rock drills"—preparing for a maneuver by clearing a patch of dirt on the ground and creating a kind of map using rocks to represent terrain features and the positions/movements of various units or individual soldiers. All these performers had a sense that doing so helped them feel a little more (maybe a lot more) prepared for their upcoming challenge, but they didn't necessarily know why the practice was beneficial, and they probably didn't get the most out of it because they were only using the visual element of their imagination instead of incorporating all their senses into the experience. Recent advances in magnetic resonance imaging (MRI) technology and brain monitoring technology (using sensitive electrodes to detect the firing of brain cells) have pulled back the curtain to reveal just how powerful the human imagination can be when used properly, and why it can help win the First Victory.

Envisioning is a confidence-building skill based on a simple but striking biological fact—your imagination stimulates actual physical changes in your body at many levels, from entire systems (cardiovascular, digestive, endocrine, etc.) to specific organs and muscles, and very importantly, to neural pathways in the brain and spinal cord which control movement and behavior. As psychologist Jeanne Achterberg wrote in the introduction to her classic book *Imagery in Healing*: "Imagery, or the stuff of the imagination, affects the body intimately on both seemingly mundane and profound levels. Memories of a lover's scent call forth the biology of emotion. The mental rehearsal of a sales presentation or a marathon race evokes muscular change and more: blood pressure goes up, brain waves change, and sweat glands become more active." In other words, your imagination is not some passive series of slide shows and movie clips, meaningless pictures that flash before your eyes then disappear with no consequence. Quite the opposite is true; your imagination, whether you realize it or not, is exerting a powerful influence on every system, organ, tissue, and cell in your body every time you use it.

Achterberg's previous statement is supported by research going back to 1929 citing the effects of imagery/visualization on everything from muscle activation to gastrointestinal activity, to immune system function. In one of the earliest investigations, Edmund Jacobson at the University of Chicago observed the electrical activity that produced small contractions in the thigh and calf muscles of a trained sprinter who was lying on a table imagining that he was running a 100 m dash. Without actually moving, the athlete was engaging the nervous system pathways that tell his running muscles to contract simply by thinking about running. While the strength of the contractions was low (the athlete didn't jump off the table), they occurred in the same temporal sequence (alternate firing of extensor and flexor muscles) that occurs during actual running. Vivid imagination apparently produces muscular effects by engaging many of the same neural structures and pathways that are engaged in the actual activity. Think of this not as some voodoo-like "mind over matter," but literally as *mind becoming matter.* This effect of neural networks being activated by the imagination has been cited in over 230 studies since Jacobson's original investigation. In one recent study Kai Miller and his colleagues at the University of Washington found that when subjects imagined simple movements such as opening and closing their fingers into fists, their brains' motor cortex area produced roughly 25 percent of the electrical activity produced during the actual movement.

The implication here is that there is a very real effect of the proper use of the imagination on the execution of any movement skill, any motor skill, or any skilled behavior. The boxer and the pianist, despite the differences in their respective professions, can both improve their skills by envisioning the correct execution, for example, scales and runs for the pianist, and combinations of jab, cross, uppercut, and hook for the boxer. The same holds true for the actor, the musician, the surgeon, the salesman, and the human resource manager. The neural pathways controlling each skill and each behavior are activated by images, especially when those images are accompanied by the sensations of related sounds and feelings. The neural pathways descend from the cortex, down through the spinal cord, and out to the specific muscles informing them when and how forcefully to contract. Each time that neural pathway is activated, the transmission becomes smoother and faster, whether you are physically practicing on the field or mentally practicing while sitting in a chair. In both cases, the neural pathways that make your movements and skills possible are being activated. Each repetition, be it physical or mental, creates a similar transmission along the existing neural pathways, and each transmission, as we noted in Chapter One, lays down another layer of the myelin sheath to allow smoother, faster, more coordinated execution in the future. Surprisingly and importantly, at a very meaningful level, the human nervous system, that incredibly complex network of neurons, synapses, and chemical transmitters, does not distinguish real stimuli (actual physical reps) from imagined stimuli (mental reps), as long as those imagined stimuli are sufficiently powerful.

What makes envisioning an even more powerful process is the fact that your muscles are not the only part of your physiology that responds to your imagination. In the last five decades a large body of medical literature has shown that imagery produces changes in physiological processes such as blood glucose levels, gastrointestinal activity, heart rate, and immune system function. Imagery practice has significantly increased the white blood cell count of cancer patients (important!), reduced postoperative pain levels, and increased the production of immunoglobin A, a class of antibodies naturally occurring in the secretions that line your gastrointestinal, respiratory, and genitourinary tracts, serving as the first line of defense against attacks by invading microorganisms. Not only can the proper use of the imagination make you more skilled, it can also make you healthier.

If you're still the least bit skeptical that the pictures you create in your imagination produce changes in your body, take this test with me:

> Imagine sitting at your kitchen table . . . picture that kitchen, the color of the walls, the position of the windows, and the table you are sitting at . . . now picture a small plate in front of you, upon which is a ripe, bright yellow lemon . . . you can see the lemon clearly, the way the light reflects off its waxy surface, the tiny bumps and dimples on the skin . . . pick that lemon up and feel its weight and texture . . . put it back on the plate and notice the slight residue it left on your fingers and palm . . . return your gaze to the plate with the lemon on it and notice that next to the plate is a small, sharp knife, the perfect kind of knife for cutting small fruit . . . Pick up that knife and carefully cut the lemon in half, feeling the blade cut through the skin, and then the center . . . put down the knife and pick up one of the halves . . . you feel its weight, lighter than before, and

> its texture, softer and squishier than before . . . you easily squeeze it gently and see some juice form on the cut surface . . . bring that half lemon up toward your face and take in that unique scent . . . you can feel a little of the juice leaking out of the lemon onto your fingertips, slightly sticky . . . now bring that half lemon all the way up to your mouth, putting it to your lips and tasting it gently . . . now take a generous bite of the lemon and really experience its full bitter taste and juicy texture . . .

Assuming your imagination works like most people's, and assuming you've handled a lemon before, the following things happened when you imagined biting into that imaginary lemon: your nostrils flared, the muscles around your mouth tightened, and you produced some saliva. Why? Because the elements of your autonomic nervous system, which (1) control the action of your facial muscles, (2) initiate digestion, and (3) allow you to taste what is in your mouth, all operated in the absence of an actual lemon. The taste receptors on your taste buds on your tongue sent messages to the gustatory cortex of your brain, telling it that something bitter was entering your mouth even though there wasn't anything to taste. Your brain sent a return message to the nasalis muscle in your nose telling it to contract and close off your nasal passages in response to a sudden strong odor even though there was no odor to smell. And your seventh and ninth cranial nerves sent messages to your parotid, submandibular, and sublingual glands to produce some saliva to begin the digestion of lemon juice and lemon pulp even though there was no juice or pulp to digest. Congratulations, you just fooled your nervous system by envisioning—creating a vivid, multisensory imaginative experience. And you can take advantage of this biological fact to make yourself more skillful at any activity you choose. Think about that.

As interesting and useful as this phenomenon may be for improving skill levels, I believe the greatest value of envisioning lies in what it can do for your confidence, particularly your sense of certainty about your future. Just as the science supports the effects of movement imagery on movement skills and the effects of different forms of healing imagery on healing processes, the science also supports the positive effects of proper envisioning on one's sense of self, the ideas you have about who you are and what kind of future you can have. Vividly picturing yourself acting in a desired fashion and accomplishing desired achievements, the science tells us, alters your self-image, that dominant opinion you have of yourself. That self-image is the beginning and end of another self-fulfilling prophecy, the kind we explored in Chapter Three. That self-image provides your powerful unconscious mind with a mental destination that you'll be motivated toward, a virtual set of instructions, like a construction blueprint, for what you might achieve. Jodie Harlowe and her colleagues at the University of Southampton found that individuals with eating disorders scored significantly higher on self-concept and self-image inventories when they created and held a positive image of themselves "as vividly as possible." They also found a similar effect in the negative direction, when the participants were instructed to hold a negative image of being "sad, upset and stressed out." These findings and those of many similar studies in clinical psychology and performance psychology tell us that the images we hold strongly influence how we feel about ourselves now in the present and about how we feel about our future. Deliberately controlling these images, because of the way they create changes in your brain and body, can be thought of as a way to exercise the muscles of either hope or despair. As Bernie Siegel, MD, put it, after years of practicing cancer surgery and discovering how powerfully a patient's attitude affected their healing, "The emotional environment we create within our bodies can activate mechanisms of destruction or repair." The images we choose to create, maintain, and energize either build up our mental bank account by presenting optimistic content and creating a biochemistry of certainty, or lower our bank account balance and leaving us hesitant and worried about our future.

So when Mike Powell was envisioning the successful achievement of his track and field record he was creating that constructive biochemistry and that constructive self-fulfilling prophecy. Same for Paul Tocci envisioning the success of his fledging business venture, and same for Colonel Capra's company commanders envisioning their ideal training. This kind of envisioning, where you create your own virtual reality of how you want to be when you step into the office, or onto the court, or up onto the stage, or wherever your personal battlefield may be, is what sets you up for breakthrough moments. This is a mode of thinking that supplies the confidence for the high school athlete to make the transition to college play, for the student to make the breakthrough to Dean's List, for the middle manager to make the breakthrough to vice president and beyond. In so many cases like these, the athlete, the student, and the middle manager all have the required physical, technical, and tactical skills, all the necessary "know-how" and abilities, but lack that element of confidence, the certainty about themselves that will allow the breakthrough to happen. They do not totally "see themselves" playing successfully at that level, or earning that recognition, and hence they feel uncomfortable. Moving up to a new and somewhat unfamiliar level of competition or responsibility, whether it is playing at a national tournament, competing in the Olympics, or taking on that big new sales territory, can create just enough worry, discomfort, and tension to undermine your well-developed and perfectly suitable skills.

A useful way to ensure that your hard-earned skills don't fall victim to fear, doubt, and worry is to *create a déjà vu experience of that desired breakthrough*, by envisioning so completely and so vividly how you want that breakthrough to unfold and how you want to feel when it happens, that when the opportunity to reach that new level presents itself, when you do get on that field or that court or into that conference room, you feel like you've already done it before and done it very well. This was precisely the experience of Olympic gold medalist Sylvie Bernier at the 1984 Olympics: "I knew I was going to dive August 6 at 4 o'clock in the afternoon for the finals. I knew where the scoreboard was going to be on my left, and I knew where the coaches were going to be seated. Everything was in my head. I could see my dives exactly how I wanted them to be: perfect dives. And when I got on the medal podium it was like I had been there before."

This is the experience I want you to have in the key performance moments of your life, the comfort and security that comes from having been there, done that, and done it exceptionally well. You can have that feeling, that utter sense of certainty about yourself, by repeatedly becoming the central character in a movie, complete with surround sound and incredible special effects. As the central character in this movie you execute your various tasks and skills at a near-perfect, breakthrough level, achieving a desired outcome without hesitation, doubt, or analysis. In this movie, or better yet, in this "virtual reality" that you immerse yourself into, you stay confident and free no matter what, easily letting go of any error or imperfection. In this high-definition, multisensory reality you joyously go for your dreams and leave the rest of the world behind. Ready?

Creating Your Breakout Déjà Vu Experience

Part One: Warming Up Your Envisioning "Muscles"

Let's begin with some warm-up exercises for your imagination. These will prepare your natural abilities of daydreaming and fantasizing for the more detailed and specific envisioning episodes that "fool" your nervous system, create physical changes in your body, and lead to personal breakthroughs

First, go back to the "lemon exercise" from earlier in this chapter. Follow those instructions once again and allow your imagination to see that familiar object in that familiar space, but this time let your mind create some more details.

- What color is the plate upon which the lemon is sitting?
- Which hand do you pick it up with?
- What subtle sounds are created as your knife slices through the lemon?
- Is that knife all metal, or does it have a wooden or plastic handle?

Take your time to create these details. Think of yourself as the producer/director of a movie scene and you have an unlimited budget to create the perfect setting for that scene. You may find that some of your imagined "senses" are stronger than others, that it's easy to envision colors and shapes, but harder to produce sounds or smells. This is normal; everyone has their own set of internal sensory preferences. You will find, however, that with a little practice your ability to create those sounds, smells, or other sensations will dramatically improve.

Now try this: Select an item or tool that you work with in your chosen profession, field of study, or sport. It could be the tennis racket, the lacrosse/hockey stick, the football/basketball/baseball, or even the particular type of shoe you wear as an athlete. It could be the instrument for the musician, the scalpel for the surgeon, the desk phone or cell phone for those whose performance arena is the office. You're going to envision this item and manipulate it mentally, warming up your ability to control any image you generate. This will expand your role as producer/director of a movie scene to include the role of special effects coordinator, the person who makes the movie scenes dynamic and memorable. Here we go. (Whenever you come to . . . pause for a count of "one one thousand, two one thousand, three one thousand" to allow yourself a couple seconds to create the details in your mind.)

> Allow yourself to create an image of that item or tool in your mind's eye . . . Imagine it suspended in the air before you, against a gray, nondescript background . . . Take in the item's color and shape . . . Zoom in on one of its edges and mentally trace its outlines, going around it clockwise, and then again counterclockwise . . . Now gently rotate that item in your imagination so that you

see its side . . . then its back . . . then its other side . . . and finally its original position . . . Reverse the direction of rotation and see the side . . . the back . . . the other side . . . and the front again . . . Now let yourself have a little fun by imagining that item gently flipping or spinning forward . . . and then backward . . . Mentally control the speed and direction of the item's movement, as if you had some superhero's ability to telepathically move objects by simply concentrating on them . . . Now, using that same superhero power, mentally bring that item into your hand (or both hands depending on how big it is) and feel it in your hand(s) . . . Feel its weight and shape in your hand as you continue to see it clearly . . . Run your hands over its entire surface so you experience all the different tactile sensations it offers, noticing how different its various parts feel to the touch, (e.g., the grip, the frame, and the strings of the tennis racket) . . . Now manipulate that item a little; pass it from hand to hand or toss it and catch it . . . feeling it as it moves and also feeling your arms and hands move as you control it (careful with the scalpel, please!) . . . Now use that item for its intended purpose; throw or shoot the ball, cut with the scalpel (carefully!), play a scale or tune on the instrument, make a call on the phone, feeling the item in your hands and feeling the motions that accompany its use . . . Now let the image fade and come fully back to this page.

What you have just done is experience and strengthen your ability to *control* what you envision and amplify the effectiveness of those controlled images with specific *detail*. These two points, control and detail, are essential characteristics of effective envisioning, and you will need to apply them as you go forward into the next exercises where you will be getting mental reps and then envisioning a great performance. As long as you control the content of your envisioned images, limiting them to scenes of success and progress, you can make valuable deposits into your mental bank account. Conversely, if you permit images of failure or difficulty to remain in your mind's eye, you draw the bank account down, subtly but powerfully strengthening the nervous system pathways associated those difficulties. Control of your envisioned content is essential. If at any time, during any envisioning work that you do, should a picture of a mistake or poor performance flash before you, do what any movie director would do and holler "Cut!" to stop the scene from going forward any further. Then, just like that movie director, immediately reset the scene and run it again so that it finishes with the ending you want. In your imagination you can have all the success you've ever wanted, achieve anything you've ever desired, and defeat any opponent or adversary, even the ones you've never beaten before. In fact, before you can beat those opponents on any field or court, you must beat them in your mind. Keep your envisioning positive!

Just as essential is the level of detail in your envisioning. The greater level of detail you create, the greater the number of neural pathways you engage, and hence the more completely you "fool" your nervous system and strengthen the pathways that control successful execution. Achieving this high level of detail simply means engaging the maximum number of senses and dialing each one up to their highest intensity. This starts, obviously enough, with maxing the visual details in any imagined scene. Where is it taking place? Indoors? Outdoors? Which room or rooms? If it's indoors, what do the walls, floors, and ceilings look like? If it's outdoors, what do the ground, the surroundings, and the sky look like? Can you give the scene exact colors and "see" the position of any equipment, furniture, tools, teammates, opponents, coworkers? Remember, you are the producer of this movie scene so you can add any details to make it realistic. How about sounds? Is there the background hum of a crowd or audience? Are there vocal cues and commands from teammates or coworkers? Announcements from a PA system? Add these audio details to accompany your sharp and clear pictures.

Now take your detailing one step further and add in other sensory specifics—the temperature of the air, the precise feel of your clothing or uniform on your body, that familiar sensation of the instrument you play or the stick/racket/ball you use in your hands, the microphone and podium you speak from. Create a multisensory virtual reality experience for yourself. With this enhanced degree of control and detail you can now move to more complex, more sequential, and more breakthrough-related mental images.

Part Two: Practicing to Improve

With your envisioning muscles now warmed up, let's give them a little workout that will help you improve a skill that you'll be using in an upcoming performance. For this exercise, select a technical skill from your sport, or a task you perform in your workday that you would like to improve. Select something that you know you'll need in an upcoming game or test, or one that is commonly evaluated in your work and one that takes a relatively short time to complete. It could be anything from hitting a more powerful tennis serve (as opposed to playing an entire match), to playing a smoother short piece of music on an instrument (as opposed to playing the entire Third Brandenburg Concerto), to entering data faster into a database (as opposed to completing your entire monthly expense report). Envisioning the execution of this skill or task will initiate activity in the brain's motor cortex and activate the nervous system pathways that control the task, due to the fact that your nervous system doesn't distinguish very well between something you actually do and something you vividly imagine doing. With those nervous system pathways activated from motor cortex to spinal cord to your body's periphery where the nerves connect up with your muscles, you might feel a slight twitch as you envision moving.

You'll be following the example of former NCAA and Olympic alternate hammer thrower Jerry Ingalls, who made dramatic improvements in his skills by diligently envisioning with clarity, control, and focused feeling. Ingalls, now a pastor in Indiana, was a skinny (six four and 180 pounds) seventeen-year-old and had never even seen the hammer throw when he arrived at West Point as a freshman. Four years later, after taking up the hammer throw on a dare by his roommate, he graduated having set the West Point record (which still stands today), having won the Patriot League Champion and IC4A championships, and having qualified for the 1996 Olympic Trials. The hammer throw is one of four Olympic throwing events, requiring very precise coordination, very precise timing, and great physical strength, to swing a sixteen-pound iron ball connected to a hand grip by a wire through four complete turns and launch it for maximum distance. Jerry Ingalls, starting as an absolute beginner, had a lot of complex technique to learn and refine. After graduation from West Point and then from Ranger school (where he broke his leg and lost thirty pounds), Jerry was selected for the US Army's World Class Athlete Program (WCAP) and returned to West Point in 1998 to train for the 2000 Olympic Trials. Once again, he had a lot of technique to relearn and refine if he was going to compete on the world's biggest stage.

To improve his skill level in the highly technical four-turn hammer throw, and to mentally prepare for competitions, Jerry became a regular client of mine during both his cadet years and his WCAP pre-Olympic years. At least twice a week in my office, and daily on his own, Jerry would get dozens of extra mental reps of his throw, firing the neural pathways that governed the movement of his feet, his knees, his hips, his shoulders, elbows, and hands, all without further exhausting his muscles or straining his joints. He would envision very specific facets of his technique, like maintaining proper head position and relaxed shoulders, or accelerating the ball from the low point of each turn through to the highest point of the turn while maintaining a straight back and deeply bent knees and ankles ("double foot catch to acceleration to single foot support hang to catch" in hammer throw geek language). He would also envision performing his entire throw, from walking into the circle to departing the circle, from first motion to final release, all envisioned in "real time," and all envisioned with the desired rhythm, flow, and feel.

Ingalls's example illustrates two important considerations when using envisioning, especially for skill improvement. The first is consistency, making envisioning a regular part of your practice routine or daily routine. Just as going to the weight room twice a month will do little to nothing for your physical strength, doing skill-specific imagery "once in a while" will do little for your skill development. Jerry felt the fifteen minutes he devoted to his skill imagery five a days a week was a small price to pay to get the benefit of a few hundred extra reps, especially when each rep was executed perfectly (he exercised control), and getting all those perfect reps put no further strain on his body. Consider the effect a few hundred extra reps a week of an important skill or key behavior might have on your ability to deliver the goods when it counts. Is fifteen minutes a day worth it?

Jerry's story also illustrates the importance of doing your envisioning from the proper *perspective*, the position from which you "see" your generated images. To make each envisioned throw activate the maximum number of neural pathways, I had Jerry envision from within his own body, using what's known as an *internal perspective*. When he was getting his mental reps, he "saw" his hands extended out before him in his starting position as he "felt" his wrists, hands, and shoulders begin their motions to start the hammer moving. He "saw" the hammer fly into the distance as he "felt" his hands release it at the end of the throw, precisely as he would have if he had actually been physically making it.

Envisioning from this internal perspective, where you are looking out from within your own body (think GoPro) and seeing what you see when you are actually performing your skill, creates a more powerful overall mental and emotional experience than envisioning from what is known as the external perspective, where you look back at yourself from outside of your body, the way you would if you were watching a video of yourself. The internal perspective is more powerful in two ways. First, it creates a greater sensation of your body in action, moving through the imagined environment, and second, it helps brings some genuine emotion to the imagined scenes. While some research suggests that the external perspective may be useful for beginners in learning a new skill (we learn by watching and then imitating, right?), the science from a 2016 review in the *Journal of Sports Science and Medicine* is also clear that envisioning from the internal perspective produces a higher level of muscular activation and is more effective when rehearsing a skill you can already perform but wish to perfect. It's not clear from Jacobson's 1929 research paper whether his sprinter was envisioning from the internal perspective when lying on the table with the electrodes attached, but I'd bet money on it.

To appreciate how important this difference between the internal and external perspectives are, imagine for a moment that you are standing in the parking lot of an amusement park watching the roller coaster from a distance. You can see a train of cars climbing up to the coaster's highest point and then swooping down the track and swerving around a curve. Perhaps you can even hear a few squeals and screams from the riders in those cars. Now imagine that you are seated in the very front seat of that roller-coaster car and are looking forward as it climbs up the track toward that very same high point. You get to the top of the track and then your car suddenly dips forward, and now it's speeding down the track so fast your heart almost leaps out of your mouth. Which of those two scenes "felt" more intense? Watching the coaster from the parking lot or being in the very front car? My guess is that your heart beat a little faster, your blood pressure increased a bit, and a few muscles twitched as you imagined speeding down that hill from the perspective of being in the very front seat. This difference in perspective, shifting from the external to the internal, changed the degree of realism that you felt and brought more of your senses into play, making your envisioning a more complete personal virtual reality experience and, in so doing, "fooling" your nervous system a little more.

In addition to promoting a greater degree of physical or kinesthetic feeling, executing his mental reps from the internal perspective where he was fully in his body allowed Jerry Ingalls to experience more of the same emotions that accompany his physical practice—pride in his execution, engagement in the moment, urgency and determination. Genuine emotion plays an important part in making your envisioning truly effective. Like any other form of practice, you can't be half-hearted about it if you expect it to work. Just as it's important to dial up the intensity of your sensory details, what you see, hear, and feel physically, it's important to dial up some emotional intensity when you envision. Think of "emotional content," as another sense to engage, right along with visual details and feelings of movement. You want to be as totally present and engaged in your envisioned scene as you are when you have a vivid nightmare. Nightmares are perhaps the best example of how the nervous system doesn't distinguish between something real and something imaginary. When you have one, your entire body responds, your heart beats faster and your muscles tense, all because the emotional aspect of the dream is so strong. It is this experience of feeling, generating powerful sensations of movement and emotion, that separates effective envisioning from everyday daydreaming and common "visualization." It's the elation that long jumper Mike Powell felt each time he envisioned breaking that world record. "I could actually feel it," he said, "feel the rush in my head." Powerful feelings like that are valuable deposits into the mental bank account. Bring that level of genuine intensity to your envisioning practice and your nervous system will respond with improvements.

Try it right now. Use the guidance below to get a few quality "mental reps" of the skill or task you've selected.

> Create the mental picture of the place where you'd be practicing your selected skill or executing this particular task (the tennis court, the practice room, your desk at work, etc.) . . . Using your ability to control your images and fill in sensory details, "see" in your mind's eye what you see when you are actually in that physical environment and positioned to execute that task (e.g., standing, seating, etc.) . . . Give your image more detail by adding sound, "hearing" what you hear when you are in that place . . . Complete the image by feeling the different feelings that you experience when you are in that place . . . Start with feeling your starting position, the sensation of your feet on the floor, court, or turf . . . Add in the sensation of any tool or object you might have in your hands . . . Sense also the temperature of the air and any smells that are present in that space . . . Complete your scene by feeling a sense of determination and purpose to improve your execution of the skill or task you've chosen . . . Why are you practicing? How is this skill or task important? . . . Allow yourself to feel proud that you are making the deliberate effort to practice and improve . . .

> Now, with the physical setting, your starting position, and your intention firmly in your mind, imagine what it might feel like to execute that task effortlessly and effectively, better than you ever have before . . . Feel your hands, limbs, or even your entire body move through the task to its completion . . . Once it's complete, hold on to the feeling of success for a moment. Experience just a little of that "I'm getting it" feeling that you had (hopefully) years ago as you were learning to pedal and steer your two-wheel bicycle for the first time . . .
>
> Take three more mental reps, carefully controlling each one so that each is executed beautifully, and so that each is full of rich details . . . allow yourself to feel a little satisfaction with each completed rep . . . At the conclusion of your third rep, take a breath, and let the image of that practice scene fade . . . allow your eyes to gently open and come back to this page . . .

Depending on what you chose to envision, a single repetition might take only a few seconds (e.g., a tennis serve). If that's the case for you, you can mentally reset yourself at the end of your rep and repeat the process for ten or twenty reps. If your task is a longer one (e.g., a dance routine, assembling a complex piece of equipment, or a presentation you'd like to refine), just one or two reps may be enough to start with. It's the quality of your envisioning—the clarity, the control, and the focused feeling that matters—rather than any arbitrary number of repetitions.

Part Three: Nailing a Great Performance

Cadet Dan Browne eases himself into the ergonomic recliner in my West Point office. Nicknamed the "Egg Chair" because of its resemblance to a large egg propped up on one edge, and with an opening that allows one to sit back into a comfortable cushioned interior, this recliner, whose actual name is the Alpha Chamber, is where Dan has come to win a particular First Victory. This is where he will set a new West Point record for the mile run and become the first cadet to run the mile in less than four minutes. I let him settle in with a few minutes of guided breath control and muscular relaxation, letting the feet, then the legs, then the arms, and finally the face release any tension. Then we mentally travel to West Point's Gillis Field House, where in two days Dan will race and attempt to step into the "sub-4" galaxy, first experienced by Roger Bannister in 1954 and considered the benchmark of a truly elite miler ever since. We picture the field house and the track in vivid detail, noting the precise way the sounds of a track meet echo off the cavernous walls and ceilings. We mentally experience feeling that unique mixture of adrenaline and anticipation that only race day can produce. Dan follows my narration of his warm-up routine, imagining the various jogging, striding, and stretching that he will perform. He follows my voice as I guide his imagination right up to the starting line for his race, feeling his spikes bite into the track surface as he crouches in anticipation of the starting gun. "Racers take your mark . . . set . . . BANG!" I start my stopwatch as Dan is off, vividly envisioning each stride and each turn of each lap of the entire race. The heart rate sensor on his finger picks up the surge he's feeling as he imagines holding his pace for nearly four minutes and then kicking for the finish line. He mentally crosses the finish line well ahead of the other racers and raises a finger, eyes still closed, the signal for me to hit the stopwatch. It reads 3:59.7. Two days later, in the actual race, Dan Browne comes in first having run that exact same time.

Canadian tennis star Bianca Andreescu was not supposed to beat Serena Williams to win the US Open tennis championship in September 2019. Andreescu started the year ranked 152nd in the world, and Williams had a history of playing especially well for the fans at the National Tennis Center in New York (she's won that tournament six times). But Andreescu was the better player in that match, and like Dan Browne, she had prepared herself to defeat Williams, perhaps the greatest female tennis player of all time, through both diligent physical practice and diligent visualizations, right up to morning of her championship match. As reported by the Canadian Press news agency, "Bianca Andreescu began her day Saturday the same way she started every morning during her run to the U.S. Open title. By meditating and visualizing how she could beat her next opponent. And Saturday's visualization session—where she saw herself defeating American superstar Serena Williams for the U.S. Open championship—worked especially well. 'I put myself in situations (that) I think can happen in a match, basically,' Andreescu said Saturday night, hours after downing Williams 6-3, 7-5 in a thrilling women's final at Arthur Ashe Stadium. 'I just find ways to deal with that so I'm prepared for anything that comes my way. I think your biggest weapon is to be as prepared as you can. I really think that working your mind (is important) because at this level everyone knows how to play tennis. The thing that separates the best from the rest is just the mindset. I guess that visualization really works!'"

As these stories reveal, Browne's record-setting victory in his mile race and Andreescu's upset victory over Serena Williams came *after* they had both envisioned it happening. *Before* they stepped into their respective arenas, Browne and Andreescu had already experienced their breakthroughs.

Here is a blueprint that you can follow to do the same for yourself, using all the key elements of envisioning that you've learned—complete control, maximum sensory details, the internal perspective, and genuine emotion—all executed from the comfort and security of your private space. Here we go, step by step.

Begin with a clear sense of your desired outcome. What accomplishment, what achievement, would bring you a tremendous sense of satisfaction? Quite simply, what's your dream? What idea, when it flashes before you, produces a physical reaction, like a little shiver up your spine and makes you say to yourself "that would be soooo great"? This could be a long-term dream of yours, like Bianca Andreescu's dream of playing (and winning!) a Grand Slam tennis final, or your next significant performance milestone, as was the case with Dan Browne's sub-four-minute mile and Paul Tocci's multimillion-dollar buyout of his fledging business. Whatever you desire to achieve at this time in your career or experience at this time in your life is right for this exercise, even if it's something that seems improbable and a little beyond your grasp right now. In fact, it *should* be something that's above and beyond what you do at present, but something that excites you to think about. As I ask all my clients, "What gives you goose bumps?" Take whatever time you need to answer that question, jot down the answer in a notebook or in the margin of this page, and get ready to envision it. Remember that you are producer, director, camera operator, and special effects coordinator of the greatest movie studio in history.

Next, set the scene for your breakthrough. Where will it take place? If you have been in that place or setting before, can you picture it in crystal clear detail? If you've not been to that place or setting before, can you find photos or video clips of it to help you create a crystal clear image of it? For athletes training for a particular event, from Super Bowl and Olympics to county championship, online photos or virtual tours of the stadium, the pool, track or field will do the trick. For the musician or dancer it's the concert hall, for the sales pro or manager it's the conference room. Find whatever representation of your performance arena you can, so that when it's time to create your multisensory virtual reality experience of success in that setting, you have some good images to start with.

Last thing before you start: list out a few of the key moments you expect to encounter in your breakthrough performance, the moments that are likely to have real impact on your success. I have found that approximately twenty minutes is the longest period of time that most people can sustain the control, detail, and emotion needed for envisioning to effectively "fool" the nervous system. If your intended performance is a short one, like Dan Browne's four-minute mile, you can envision it in its entirety, stride by stride or second by second, in real time. But if it's a longer event, like Bianca Andreescu's three-set championship tennis match, you won't be able to envision it from start to finish. For these longer events, like recitals, presentations, surgeries, or cross-examinations, it helps to pick out the moments that are particularly important, and envision each of these moments in the order they occur, with full control, vivid detail, and genuine emotion within that twenty-minute window of time. Certainly the start and the finish of your event are such key moments. Others might be the tougher passages of the music or dance performance, the more challenging parts of the surgery, the moments in your presentation where you are communicating a particularly important concept. Once you have these moments identified, it's time to zero in and put your imagination to work.

We will begin this envisioning practice by setting up a personal workspace. I don't mean a room in your home or office, but instead a place in your mind where you go when it's time to envision for skill improvement and breakthrough performances. This private room is your safe zone, your haven, your sanctuary, and when you're in it, everything and anything is possible. Make sure you are sitting comfortably with your back supported. You may want to have a friend read these next paragraphs to you so you can envision more easily with closed or softened eyes. As before, whenever you come to ". . ." pause for a count of "one one thousand, two one thousand, three one thousand" to allow yourself a couple seconds to create the details in your mind.

Step 1: Create Your Private Room

Imagine you are walking down a well-lit, carpeted corridor with doors on either side, the kind you'd see in a hotel . . . At the end of the corridor, directly in front of you, is a door with a special symbol on it, a symbol that identifies this as the door to your private room, your personal mental practice space . . .

Open that door and step through it into a room that you have created for just this purpose. It is furnished and decorated exactly the way you want it . . . Take a moment to look around this room and see the walls, the floor, the ceiling, and everything that you have put in this room to make it your very own; paintings, photographs or posters that you love . . . plants, sculptures, and any other decorations that make this place truly special for you . . . Perhaps there is a favorite piece of music playing in the background . . . The windows in this room (if you decided to have some), look out on a landscape of your choice—perhaps a warm sunlit beach, or a tranquil lake high in the mountains, or a beautiful garden blooming with bright flowers, or the bright lights of a big city, whatever brings you the greatest sense of contentment and belonging . . .

Move through your private room over to a comfortable chair or recliner that you have placed exactly in the just the right spot . . . Sit down in that chair and feel your body settle into a comfortable but supported position as you take in the details of your room . . . Next to your chair is a small table, and on that table is your favorite drink in your favorite glass or mug . . . Reach out and bring that drink to your mouth, taking a satisfying taste . . . Now place it back on the table and settle into your seat, taking a few easy deep breaths . . . Let your face and jaw relax as you exhale, knowing that here in your private room you are safe and secure, that you have all the time you need to get perfect mental reps of a skill or skills, or envision your next breakthrough performance . . .

Step 2: Arrive at Your Arena

Once comfortably settled in your private room, let your imagination take you to the starting point of your performance. If you have been to this particular arena before, use your memory and the internal perspective to imagine really being in that place. If your breakthrough is to take place somewhere that you haven't been yet, use the photos or videos you've found to set the scene, also using that internal perspective . . . "See" from out of your own eyes the door to the locker room where you change into your competition uniform, or to the building where you will be giving your sales presentation, or to the concert hall where you will perform your recital . . . Every performance has a starting point from which there is no turning back. Envision that starting point in all its detail, allowing the sights, the sounds, and all the associated feelings to be fully experienced . . . Envision moving from room to room or place to place in that setting, feeling your autonomic nervous system kick in with a few of those butterflies, just as nature intended it to . . . Arrive at your locker, or the ready room, or wherever you will go to make your final preparations . . . With the clarity, control, and focused feeling that you've practiced, put on your uniform or adjust your clothing, hair, etc. so you are ready to step out and take command of the moment . . .

Step 3: Gear Up/Warm Up/Step Up

If your performance involves any kind of warming up or tuning up, like Dan Browne's breakthrough mile race, envision performing a few of those actions . . . You needn't envision your entire warm-up routine, just enough of it, maybe one or two reps of each movement, to get some emotion going . . . See what you see, hear what you hear, and feel what you feel as you prepare your presentation materials, glance over your notes, or take one last look through the music you'll be playing . . .

Step 4: Get Off to the Right Start

Now it begins for real! Envision stepping up to the starting line, the podium, or whatever position your performance begins from . . . Feel whatever mixture of excitement and anticipation you are likely to feel . . . See or hear your cue to begin and immediately feel yourself take control of the action . . . Your first serve skims the service line for an ace, your first stride out of the blocks puts you in front of the pack, your first note is delivered with perfect pitch and volume . . . Spend a minute envisioning establishing your desired pace, your desired rhythm, your desired connection to your audience or your customers . . .

Step 5: Hit the Big Moments

Now use your list of important moments and envision each one with full control, clarity, and emotion . . . "See," from your internal perspective, the action unfold around you as you flawlessly execute your game plan, your presentation, your performance . . . see the teammates, competitors, or audience members who are there . . . see the scenery or background change as you move through that moment . . . hear the various sounds of that moment, especially your own voice if speaking is part of your breakthrough . . . and feel your position and your movement in each moment, each action you take as you proceed through that moment . . . feel also the emotional content and level you wish to have in each moment . . . are you fired up or cool as can be? . . . are you aggressive and physical, or sensitive and intuitive? . . . whatever ideal emotional state each moment in your breakthrough performance requires, you achieve it naturally . . . In each of these moments, each scene of your multisensory lived-in reality, you perform superbly and succeed in controlling your opponent or mastering the situation or impressing your audience . . .

Step 6: Finish with a Bang

Now envision that final moment, that last lap, that last inning, your closing pitch to the audience . . . Just as you did when you started, take this moment to see, hear, and feel with total clarity and control your last moves, the final notes played, the concluding step in the surgery . . . Feel whatever mixture of excitement, satisfaction, or relief you are likely to feel as the last seconds tick off the clock . . .

***Step** 7:* Celebrate

Wait! It's not over yet! Just like long jumper Mike Powell "never denied himself the elation" that breaking the world record would bring each time he envisioned in his living room, don't deny yourself the experience of some genuine happiness after your envisioned breakthrough. See the smiles of your teammates swarm around you in celebration, or your coworkers coming up to congratulate you after your knockout presentation . . . hear the cheers, the applause, the compliments for a job well done . . . and let yourself feel a sense of validation and fulfillment, knowing that your efforts have finally paid off . . .

Once you really let yourself enjoy the moment of your breakthrough, let that scene of that celebration in that arena fade, and return your imagination to your private room, seated in that comfortable seat, surrounded by the furnishings and decorations you have put there . . . Next to your chair is that small table with your favorite drink on it . . . Reach out take another satisfying taste of that drink . . . Now place it back on the table and stand up, looking around to enjoy the view out the windows . . . Now move through your room toward the door, again taking in the photos, paintings, and other details you've put into this room to make it your own . . . Reach the door to the hallway and pause to take one last look at your private space, knowing that it will always be here for you and that you can come back to it any time to achieve any breakthrough you desire . . . Step out into the hallway and close the door behind you . . . Allow your eyes to open and come back to this page . . .

These seven steps can be followed as a daily mental training routine. No equipment or special talent is needed to perform them properly and receive the benefit of a nervous system that has already experienced your breakthrough performance. It bears repeating that once your nervous system has repeatedly executed your vividly envisioned, emotionally genuine breakthrough performance, it will be primed and prepared to deliver that same performance with the barest minimum of conscious, analytical, potentially distracting thoughts. You will perform with maximum confidence because you will have won your First Victory.

But wait . . .

This chapter so far and the science that supports it have focused on the constructive effects of envisioning improvements and breakthroughs; creating in your imagination a detailed multisensory experience of a precise desired outcome. But let's face it—the world out there beyond your imagination doesn't always care about what you want, and no amount of "wishful thinking," even when it physically alters your nervous system, is going to change the fact that there are opponents, competitors, and a myriad of other forces working to prevent your preciously and passionately envisioned breakthrough from ever actually happening. As they say in the army—"the enemy gets a vote." Let's confront these forces and prepare to defeat these enemies with a special version of envisioning that I call the "Flat Tire" drill.

Let's say you're driving down a poorly lit street toward a destination you have never been to before. It's nighttime, but your car's (or phone's) navigation system is keeping you on course and on schedule. Oh, and it's raining. All is well until you drive around a blind corner and your front driver's-side tire hits a deep pothole that you didn't see coming. Ten seconds after that impact your tire warning light comes on and instead of driving smoothly your car is now limping along, *bump bump bump*. You know that if you drive any farther, you'll do real damage, so there's no choice but to pull over, stop, and change your flat tire. Worse, it's important that you get to your destination soon. Making a call to roadside assistance (even assuming you have such a service) and waiting for the tow truck to arrive and bail you out isn't an option. It's up to you and you alone to deal with it.

Undoubtedly you will not be pleased at that moment, but reality has thrown a monkey wrench into your schedule and you have to respond. How well you deal with this situation, how easily and quickly you change that flat tire and get back on the road will depend on whether you've changed a flat tire on that car before. If you already know the steps to take and know where the jack, the spare, and all the tools are, then it's still an unexpected problem, but a relatively simple one to fix. Conversely, if you've *never fixed a flat on that car before*, then you're in for a much longer and much tougher tire-changing procedure—extracting the owner's manual from the glovebox, finding the instructions and reading them in the dim cabin lighting, finding the jack, the spare, the tools, and then following the instructions in the dark and in the rain before you are safely back on the road.

We all encounter various "flat tires" as we go about our business, and it's likely that despite our best intentions, our perfectly envisioned performance breakthrough will not go off without a hitch. To build up my clients' confidence in their ability to overcome these unexpected difficulties, I teach them how to win a series of mini First Victories before they enter their respective arenas. First, they honestly identify a few likely "flat tires," and then they deliberately envision a successful response to each, establishing in their nervous system an effective "subroutine" just in case it's needed. That way, just like the motorist who knows what to do on that dark, rainy road, they can continue performing with full confidence.

Jerry Ingalls, the hammer thrower we met earlier in this chapter, encountered one such flat tire at his 2000 Olympic Trials and dealt with it admirably. He made it through the first two rounds of competition at the Trials, becoming one of eight finalists. Each of these finalists would throw three times, and the three men with the longest single throw in this final round would become Olympic Team members. Jerry entered the packed-to-the-rafters stadium at Cal State University Sacramento, strutting his 265 rock-solid pounds, and began his stretching routine. All was proceeding according to plan until Jerry stepped into the throwing circle for his final warm-up throws. There he found something totally unexpected: someone had scraped the concrete surface of the throwing circle with a stiff wire brush, creating a much rougher surface, which made for much slower turns. Apparently one finalist thought the circle was too smooth and too fast, and he had the circle doctored up a little to accommodate his preference without informing any of the other finalists. That meant Jerry had to make some last-minute adjustments to his technique right before taking the three biggest throws of his life in the biggest track meet in the world outside of the actual Olympic Games.

As you might well imagine, this sudden and surprising change of throwing conditions upset each of the other throwers—except Jerry Ingalls. As several competitors melted under the pressure of having to deal with a last-second change that they didn't think was fair, Jerry Ingalls took it all in stride, and instead of folding, he got better and better, throwing farther and farther as the final round went on. It took an amazing last throw by a veteran competitor with ten more years of international competition to knock Jerry out of third place and into the alternate slot, still an impressive achievement nonetheless for a guy in only his fourth season of competition and who just two years earlier was a 220-pound infantry platoon leader.

How was Jerry able to fix the "flat tire" that the scraped throwing circle presented? By having diligently envisioned a constructive response to many other possible problems. He and I had spent many hours in the months before that Olympic Trials meet honestly identifying the situations and settings that might stop him from feeling his ideal mix of aggressiveness and ease, the state he called "hammerama." We planned out what he'd say to himself, how he'd breathe and stretch, to get back into hammerama if he were to foul his first two throws, or if he found himself well behind the leader with only one throw remaining. Jerry would carefully envision his response to each of these, and many other situations, his envisioning always finishing up with him launching a monster throw. While he never actually envisioned being confronted by a newly scraped throwing circle at the Olympic Trials (who could have predicted that?), Jerry was so thoroughly conditioned in how to recover his composure through all his other Flat Tire drills that a surprise in the form of a roughed-up throwing surface had no effect on him at all.

For another example of a performer conditioning himself to readily handle the totally unexpected, take the case of Phil Simpson, the most successful intercollegiate wrestler in West Point history. Phil was recruited out of Nashville, Tennessee, a region not known for producing great wrestlers, but he became a four-time National Championship qualifier, a three-time all-American, and during his senior year, the National runner-up. It was during the semi-final round of that last National tournament where he overcame a totally unexpected flat tire by virtue of his previous envisioning practice and defeated a very tough opponent whom he had lost to twice before.

ESPN was carrying the action live from the Savvis Center in St. Louis, Missouri, as Phil Simpson and Dustin Manotti from Cornell were on opposite sides of the mat about to compete in their 149-pound semifinal. Like all wrestlers at this elite level, they both had well-established prematch routines to bring them to peak mental and physical readiness right before each match, and they stepped up to the edge of the mat both raring to go. They had been previously warned about possible TV time-outs between matches, so when they received a "one minute to go" warning from the ESPN producer, neither blinked an eye and both continued to bounce on their toes and shake out their arms to stay loose. But that one minute became two minutes, and as the seconds continued to tick by, Manotti's composure began to crack. As the delay reached four minutes and then five minutes, Manotti was noticeably agitated, his careful prematch routine having been meaningfully interrupted. Phil, on the other hand, spent those five minutes sitting calmly on his side of the mat thinking, *I'm in total control. I don't know when they'll call me, but I don't care.*

That all-important *sense of certainty* came easily to Phil Simpson because of all the envisioning he had previously done. Prior to this tournament and prior to many others over the years, Phil had listed out with me a series of "flat tire" situations that might turn up, and carefully envisioned his constructive and successful response to each one. He entered his competitions with a well-developed mental subroutine all built in so he was fully prepared to battle back if he was behind at the end of the first period, to change his attack strategy if his favorite takedown wasn't scoring, to remain composed if a questionable call went against him, and to tap into another gear if he needed to score some last-second points to win. Like Jerry Ingalls at the Olympic Trials, who couldn't have predicted a doctored throwing circle, Phil Simpson didn't specifically envision having to wait out a five-minute TV time-out before the biggest match of his life. But "having rehearsed all the other ones," he told me later, "I could easily handle this one. I just sat there, controlling what I could control, thinking 'this is so great!' That's when I knew all my mental work was paying off." When he finally got the go-ahead, Phil calmly walked onto the mat and wrestled to a resounding 8–0 victory to go on to the National Finals.

To execute your own Flat Tire drills, follow these simple guidelines:

List out three "flat tire" situations that might cause hesitation or doubt in an upcoming performance. Start with one that perhaps has actually happened to you in the past and then come up with more that might happen. For the executive or salesman about to make a key presentation, it could be a sudden glitch in the room's audiovisual system (who hasn't experienced that?). For the athlete, how about having one of your fellow starters leave the game due to injury and now you have to play with a second- or third-string player you haven't practiced with? For the first responder it could be any one of a dozen mechanical failures (monitor, phone, pump, etc.), or for the surgeon the most likely complications to be anticipated.

Envision the first situation on your list with full control, detail, and emotional content from the internal perspective. Make it real, make it strong. It's even okay to let yourself feel a little of the worry that your flat tire situation might bring. *But you only envision this scene for ten seconds!*

Now envision stopping that scene from going any further. Cut it off by telling yourself "STOP!" or "Time to take control!" with some real conviction.

Now envision breathing deeply and pausing just long enough, even if there is action going on around you, to relax your neck and shoulders.

Now envision, with total control, rich detail, and emotional content from the internal perspective, taking the actions that result in you returning to a greater sense of certainty in that moment. *Hear* yourself giving yourself reassurance (*I'm prepared to handle this; This is what I trained for; It'll take a lot more than this to get me off my game*). *See* the scene unfold around you as you move purposely, step by step and movement by movement, to control what you can control in that moment. *Feel* your emotions shift from worry and irritation to calmness and eagerness as you take control of the situation—the salesman putting the room at ease while the AV glitch is addressed, the athlete looking at the new addition to the lineup and saying, "Great to have you," the first responder finding an alternative to the failed device. *Envision this effective and successful response to the flat tire for at least thirty seconds. This step is the key to the Flat Tire drill. Spend at least three times as long envisioning yourself controlling the situation and getting it all get back on track as you do envisioning the problem itself.*

Finish the drill with an envisioned scene of success; the presentation is a smash hit, the new teammates makes a game-winning play, the fire/accident/encounter is controlled with minimal damage.

Conclusion—Is it "Real" or Is It "Delusion"

I am often asked by a client if the envisioning concept and the envisioning exercises are nothing more than personal delusion. "I'm not really doing anything when I imagine all these skill improvements and breakthroughs. Aren't I just fooling myself?" My answer is always the same: "Yes, you are fooling yourself. But it's precisely by fooling yourself that you make real change." As my client's brow furrows, I explain.

There was a certain rite of passage in your youth, an important transformation you went through in your early years, which at the time was rather momentous. You probably take it for granted now, but back when you were around six years old, it was a big deal. I'm referring to that important personal milestone of riding a two-wheeled bicycle without training wheels. I've yet to meet someone who accomplished this feat on their very first try. For everyone I've ever met, it was a period of trial and error, of initial frustration and some unforgiving contact with the pavement. You simply *could not do it* at first. But you had an idea that it was possible. You saw other kids do it, and Mom or Dad told you that you could do it too. Somewhere in between your first unsuccessful launch down the sidewalk or driveway, where you veered out of control and either fell or stepped off, and that great moment of exhilaration when your motor cortex assembled the right configuration of neural circuits and you pedaled away happy, you had an internal mental representation, some kind of image, of doing it right. You had a temporary "delusion," a sense that you could ride successfully despite having just fallen off. You had no actual evidence or proof that you could do it, and some immediate evidence to the contrary, yet you maintained enough of that mental representation, that delusional belief, to get back on the bike and practice pedaling, steering, and balancing until your nervous system updated its bicycling software subroutines and you "got it." Being a little bit delusional was critical to one of the happiest moments you ever had.

When you think about it, almost *every* change, every development, and every accomplishment starts out from a similar place of constructive delusion, some idea, some mental representation, that you can do something in the future or be something in the future despite never having done it or been it before and despite having some evidence that you can't.

The science on how expertise in any field is developed tells us that "deliberate practice," working right at the edge of your current ability where you are momentarily "failing," is the key. What sustains you through all that deliberate practice is a certain level of constructive delusion coupled with the proper mental filtering we've been discussing in the last two chapters.

So indeed, envisioning practice of any kind is a form of delusion, but a constructive and, dare I say, necessary version of it. The young Stefani Joanne Angelina Germanotta, now known around the world as Lady Gaga, maintained a vision of her eventual stardom long before her breakout commercial success. "I used to walk down the street like I was a f***ing star," she told *Rolling Stone* magazine for her first cover story in 2009. "I operate from a place of delusion. I want people to walk around delusional about how great they can be and then to fight so hard for it every day that the lie becomes the truth." That's good advice. Dream about your ideal future, even if it's a lie when compared to today's reality, then get to work and make that reality happen. Win that First Victory and the rest can follow.

Let's take another example, from something about as far from the glamorous world of pop music stardom as you can get—car sales. In episode 513 of the podcast *This American Life*, entitled "129 Cars," narrator and author Ira Glass takes us inside a Chrysler dealership in Levittown, New York, on Long Island to hear how car salesmen fight to make their monthly sales quotas. At the dealership we meet their top salesman, Jason Mascia, who routinely, month after month, outsells the rest of the sales crew. In a business where selling fifteen to twenty cars a month is considered a "solid performance," Jason Mascia often sells thirty and has a personal goal of hitting "forty plus." His secret? Constructive delusion. As his sales manager puts it, "Jason knows that a certain percentage of customers, depending on the month, will walk away from him without buying a car. Of course that happens. And yet, paradoxically, he enters every negotiation full in the knowledge that it will *not* happen." Jason Mascia has a vision, call it delusion if you like, of each person who walks into the dealership's door signing on the dotted line. Intellectually, realistically, he knows this won't happen, but like Lady Gaga, he operates from a state of delusion where the successful sale is a foregone conclusion that merely has to be finalized. And that sense of certainty, that First Victory, helps him seal deal after deal.

Final Note

Having learned to manage your memories of the past and your thoughts about yourself in the present, this chapter has covered how to think selectively and constructively about your future, using the mental process of envisioning. Your self-generated images have tremendous potential to energize you, to improve your physical, technical, and tactical skill sets, and to prepare you for an upcoming performance. We've covered:

1. How your mental images of both future success and future defeat create a cascade of neural impulses affecting your muscular, gastrointestinal, cardiac, and immune system function.
2. How these impulses establish neural pathways that facilitate the repetition of desired behaviors.
3. How to hack into this system by deliberately using specific imagery guidelines: the correct perspective, the incorporation of multiple senses, and the generation of genuine emotion, because the human nervous system does not distinguish real stimuli from imagined stimuli.
4. How to create and use a personal mental sanctuary, a place in the mind to safely and securely imagine successful performances in vivid detail.
5. How to use envisioning to establish a greater sense of certainty about your skills, your next breakthrough, and your ability to handle the flat tires that the bumpy road to success is sure to produce.

Armed with this new set of tools, you are better prepared to win your First Victory. The simple truth is that the more clearly you envision the accomplishment of your goals and the path to achieve them, the more likely it is that you will reach them. My hope is that you'll use these tools daily to clearly envision your next breakthrough and whatever skills or abilities you'll need to make that happen. Doing so consistently will drop numerous deposits into your mental bank account and create the habit of carrying a series of ongoing positive images with you wherever you go.

Like any set of tools, however, these envisioning techniques do you no good when they are left in the box. It could be that what is holding you back in your life as a performer is not the amount of sweat that you're willing to drop in practice, or the hours you're willing to study your craft, although we all know that doing so is necessary too. What might just make or break you is that total belief that you can achieve a given goal, a belief that is based on the fact that you've envisioned it hundreds of times, literally creating the neurology that can produce it. While there are many dedicated athletes and determined professionals in all walks of life who are willing to study or practice three, four, five hours a day, only a few, I have found, are willing to spend fifteen minutes a day vividly envisioning their most cherished dreams. That's a different kind of effort, a different kind of discipline, but it is the kind of discipline that separates the champions from the rest of the field.

CHAPTER FIVE

Protecting Your Confidence Every Day, No Matter What

How to Defend Against Threats and Outright Attacks: Locks, Alarms, and Other Antitheft Devices

Mario Barbato had just entered his office when several issues were thrown at him. As he put it in a letter to me, "A lot of negativity came right at me and problems need to be solved. I could feel my confidence shrinking and the anxiety level rising." But having learned and practiced how to handle negativity and respond constructively to the inevitable difficulties of the modern workplace, Mario had some weapons at his disposal. "I politely asked the folks in my office to step out and give me a few minutes. I sat down and went through the exercises you showed me, wrote my personal affirmations three times each, took a deep breath, and then began addressing the issues at hand." Those few moments were all he needed to win a momentary First Victory and regain, as he put it, "the right frame of mind." Doing so led to the successful resolution of the day's pressing issues, bringing Mario some unaccustomed praise from one of the senior members in his firm.

A few days later Mario was asked by his boss to coordinate a meeting with an ex-CEO of a large firm. At the last minute, the boss was called away and Mario had to step up and kick off the session himself. "I know for a fact," he reported later, "that prior to our work this type of situation would have caused me to second-guess everything I planned on doing in the meeting and dwell on the negative afterward. But instead I went back to my Top Ten and opened up my mental bank account. Then I walked into the room, went right up to the CEO, shook his hand, and helped kick off the session. I kinda felt like I let out my roar."

Mario's experience demonstrates an important reality: in order to win your First Victories in the real world and then perform confidently, you're going to have to protect your mental bank account from both actual bad events and the negative thoughts those events can produce. No matter how well you manage your memories, tell yourself the right stories, and effectively envision a successful future, life is going to hit back at you, and the precious confidence you have built up will be assaulted. The problems you encounter, the setbacks you experience, and the mistakes committed by both yourself and others will indeed enter into your running total of thoughts. Like modern cybercriminals, they can break into your mental bank account and drain it unless you have some safeguards in place. This chapter is devoted to those safeguards, the mental habits that will prevent both the external bad events (those inevitable errors, mistakes, and setbacks), and your internal negative thoughts (which *everyone* has), from diminishing the sense of certainty you want to have when it's your turn to step into the spotlight and let out your own "roar."

Safeguard #1: The Constructive Attitude Lockdown: How to Think About Errors, Mistakes, and Setbacks

Here's part of a conversation I have three or four times a week. The speaker could be a professional athlete in search of a Hall of Fame career, a high schooler shooting for the SAT scores that will help gain admission to Stanford, or a West Point cadet who wants to be the honor grad (top finisher) in his or her Army Ranger School class upon graduation.

"This is great stuff, Doc. I really like it, I know it will help me, but I've got a question." As soon as I hear this, whether it comes after a discussion of managing the memories, or of talking to oneself optimistically, or of envisioning the future, my spidey sense starts tingling; I know what's coming next. It will be some variation on the theme *What do I do when things go wrong?* Here are some samples:

- Do I just filter out all my mistakes completely?
- How should I think when I know I've played poorly?
- Should I only think about the positives in my work?
- Do you really expect me to be like the guy in the movie who ignores everything except that one-in-a-million chance?
- How am I supposed to get better if I don't think about the things I'm bad at and need to work on?

When I hear any of these questions, I know that the person asking it has only understood half of what the mental filter does and is now ready to understand the filter's full power. As mentioned previously, the filter performs *two* functions in the service of your mental bank account. The first function is allowing the thoughts that create energy, optimism, and enthusiasm to pass through and be retained. By managing your memories, using affirmations, and properly envisioning your future, you exercise this function. But the second function of the filter is just as important and it's often overlooked; *the filter also restructures, or purifies, the thoughts and memories that could create fear, doubt, and worry into helpful advice,* thus preventing them from drawing the bank account down and in fact, turning them into additional deposits.

My quick answer to the person asking any of those questions is this: "It will always be important to think about what you need to work on, and you'll always have to deal with negatives in your work or the occasional bad game. We live in an imperfect world where things do go wrong, and no matter if you are Tom Brady or Serena Williams or Bill Gates you're going to have below-average days once in a while and a below-average moment or two probably every day. So yes, you will have to think about your less-than-good moments, whenever they occur. Just be careful *how* you think about them. Here's what I mean . . ."

The second function of your mental filter is to release and/or restructure any thoughts, memories, or experiences that compromise your energy, optimism, and enthusiasm. In other words, your filter works to keep everything in a functionally constructive perspective, even when you don't have a lot to be happy or enthusiastic about. As important as it is for your filter to "let in the good," it's perhaps even more important for it to minimize the damage that the "bad" can do. Bruce Lee, the famous actor and international martial arts icon, described this in the very first paragraph of his chapter on "Attitude" in his seminal book, *The Tao of Jeet Kune Do.* The self-confident athlete, Lee wrote, who is "fed by previous successes, and *having completely rationalized previous failures* [italics added], feels himself a Triton among minnows" (Triton being a mythical god of the sea). That process of "rationalizing previous failures" is the second key function of the mental filter, and it is essential for protecting your mental bank account. The word *rationalizing* in this case does not mean denying your failures or refusing to accept the consequences of your actions. It means instead that you *explain to yourself* what happened in a way that protects your mental bank account and puts you in a position to learn, grow, and move forward. Here's how you can do that, a three-part mental method for explaining your failures and setbacks to yourself so you can keep your confidence and maybe even build a little more.

Part One

Think of each and every mistake, error, and setback as *temporary.* Yes, it happened, and it may even have been costly, but it's important to treat it like *it only happened that one time.* Thinking this way about your mistakes and imperfections prevents you from sinking into the "here I go again" trap of worry and self-doubt. If you allow yourself to think that a mistake or imperfection, once it occurs, is going to lead to more of the same, then you've opened up your mental bank account to a sneaky criminal. Instead, tie that criminal down by telling yourself "It was just that one time, and now I have a clean slate going forward." When you *treat your mistakes and imperfections as temporary*, you acknowledge them and then leave them in the past where they belong, instead of bringing them with you into the present moment and into future moments.

How might this work in a real situation? Take the case of Maddie Burns, West Point class of 2020 and goalie for the West Point women's lacrosse team. Maddie's job as goalie is to stand at the mouth of a six-foot-by-six-foot goal and get in the way of a rock-hard rubber ball that is coming at her at 70 miles per hour. She performed this unenviable task rather well in the 2020 season, finishing with the nation's fifth-best goals-against average of 7.75 (the nation's best being 7.07). That 7.75 number means that seven times in every game a shot on goal gets past Maddie and she has to turn around, dig the ball out of the goal, and hand it to the referee while the opposing team whoops, cheers, and celebrates the goal they have just scored. It's not hard to imagine how that can wear down even the most confident individual. Working with me during her junior year at West Point, Maddie took to heart the concept of treating any goal scored against her as a temporary, "just that one time" occurrence. She quickly realized that no matter how well she played, the nature of the game was that goals were going to be scored on her. That was an important acknowledgment—things were going to happen that she couldn't control, and that meant she had to respond to each of them constructively, one after another, if she was to give herself any chance of playing well. No matter how often she was scored upon, even if the opponents scored two, three, or four times in a row, Maddie's best choice was to maintain this "just that one time" attitude and protect both her own confidence and her team's. Leave your mistakes and imperfections in the past. That's where they belong.

Part Two

Think of each and every mistake, error, and setback as *limited.* Yes, it happened and, yes, it may have produced some uncomfortable feelings, but it's important to treat it like it only happened "*in that one place.*" Thinking this way about your mistakes and imperfections prevents you from sinking into the trap of thinking "this whole day is going down the drain" or "now my whole game is in trouble." The golfer who hits one of his first drives into the trees instead of onto the fairway had better keep that poor shot in its proper place instead of generalizing from that drive that now his irons, wedges, and putts, all those other parts of his game, are now in trouble. Thinking, *It was just that one drive and the rest of my game is fine* is a much more helpful and perfectly reasonable alternative. The soldier in basic training who struggles with getting her gas mask on right and then has some bad moments during her five minutes in the tear gas chamber had better keep that failure in its proper place by thinking, *Okay, I messed that up, but I'm keeping up with the pace during the morning runs, I'm doing well on the firing range, and I can fix those straps on the mask so it'll go on faster next time.* If you allow yourself to think that a mistake or imperfection, once it occurs in one specific situation, is going to affect other situations, then you've opened up your mental bank account to another insidious criminal. Arrest that particular criminal by telling yourself, *It happened in just that one place, and everything else is okay.* When you *treat your mistakes and imperfections as limited in scope,* you acknowledge them and then cordon them off, leaving them locked down in the single place where they occurred. This will help you to execute all your other tasks with more certainty. Combine *treating your mistakes and imperfections as limited* with *treating your mistakes and imperfections as temporary*, and you have a powerful one-two psychological punch—the mistake only happened that one time, and it only happened in that one specific situation—so you have every reason to take on your next task at work, your next play from scrimmage, or your next point in the tennis match with the full value of your mental bank account creating the maximum sense of certainty. First Victory won.

Part Three

Think of your various setbacks and imperfect moments as *nonrepresentative* of yourself. Once again, acknowledge that those moments happened, and acknowledge whatever damage or consequences resulted, but then choose to see those moments as inaccurate reflections of who you are and what you are capable of. Where treating your mistakes as *temporary* protects you from the "here I go again" trap, and treating your mistakes as *limited* protects you from the "my whole day is going down the drain" trap, treating your mistakes as *nonrepresentative* protects you from the "maybe I'm not good enough" trap, that swamp of unrestrained self-criticism that is always waiting to engulf us. These confidence-sucking traps can open up an ever-widening expanse of self-doubt, starting out with the feeling that a single discrete mistake will repeat itself over and over again ("here I go again"), then expanding to the feeling that mistakes will soon occur in more and more places ("now my whole day is going down the drain"), finally expanding even further to the feeling that you as a player, or a professional, or even as a person, aren't good enough and maybe shouldn't bother trying anymore. There is a curious tendency in our modern world to overidentify with our shortcomings and even define ourselves by our mistakes, presumed limitations, and all the things we can't yet do. Doing so, not surprisingly, is a confidence killer. It's also completely unnecessary; there's no law requiring you to think that way. Put that third and final devious criminal in jail by thinking, *That's not how I really am, I'm better than that, it must have been some kind of fluke*. Then throw away the key.

I witnessed a great example of this constructive response to a series of mistakes some years ago during an early season practice of the West Point men's lacrosse team. Head Coach Joe Alberici had become understandably unhappy with the team's execution of a full field passing drill that afternoon; too many balls were being dropped way too often. So he blew his coach's whistle and stopped the drill, Then, instead of berating the players for the poor quality of their execution, he stepped to the middle of the practice field and shouted "We're better than this! I'm not sure what's going on, but I'm sure this isn't us." That statement, "This isn't us!" rationalized the team's failure to execute by *externalizing* it—pushing it away from the team's collective confidence by proclaiming that the dropped passes were *not representative* of the team's ability. Coach A then took his message a step further by walking over to one of the dropped balls, scooping it up in his hand, and announcing to the team "It's gotta be this ball!" whereupon he took the lacrosse stick from the nearest player, placed the offending ball in the stick, and launched the ball high into the stadium's cheap seats. "Now let's get it right!" he shouted, returning the stick to the player. The drill resumed and the team resumed its characteristic passing efficiency.

Was it the ball's fault? Of course not, but Coach Alberici

wasn't crazy for suggesting that it was. What he did by declaring, "It's gotta be this ball" was to take some pressure off his players and reinforce the fact that he believed in them. Rather than say anything that might get them thinking, "We're stinking up the place right now," and thus engage the negative "sewer cycle" of the mind-performance interaction from Chapter One, Coach Alberici protected their confidence with both his words and his actions.

When you experience your own version of a practice drill that isn't going well, or an unexpected setback, or even a protracted downturn in your productivity at work or at school (a "slump" in sports talk), will you *personalize* that problem and let it pull you into a series of "what's wrong with me?" questions, or will you *externalize* that problem by mentally *putting it outside yourself* by thinking some variation of *That's not how I do things* or *I'm a better player than that* or even *Wow! Can't believe that just happened. There must be some weird alignment of the stars tonight*. Doing so in no way absolves you of the responsibility for your actions or of the consequences of those actions; you won't get very far in life if every time you make a mistake you shrug it off by exclaiming "the devil made me do it!" as the old comedian Flip Wilson used to say. And it doesn't mean avoiding that honest look at yourself and realizing that you might need to put in some more practice or learn some new skills. It just means, first and foremost, that you maintain a sense of your fundamental value as a person and then add to that the acceptance of whatever level of competency you've developed through your practice and study. Top that off with a level of curiosity to see just how well you can perform the very next time you meet a new client or get your next at bat or take that next shot. Armed with this new skill of treating your imperfections as nonrepresentative as well as temporary and limited, you have a one-two-three combination defense against any loss of confidence that life's inevitable setbacks will throw at you.

These methods for protecting your confidence are derived from and supported by a large body of research pioneered by the father of positive psychology, Martin Seligman, and his many colleagues. Having begun his research career studying how both animals and humans can become pessimistic and depressed even when positive alternatives exist for them, a phenomenon he called learned helplessness, Seligman turned to studying optimism and health in the 1980s when he found that some individuals stubbornly refuse to become pessimistic when confronted with troubles and difficulties. In one of his seminal books, appropriately titled *Learned Optimism*, Seligman pointed out that the optimistic individuals who resist becoming depressed differ from more pessimistic individuals not in terms of IQ, talent, or motivation, but by their "explanatory style," the way they explain to themselves the causes of both the good events and the bad events they experience. Pessimists, he noted, tend to take the view that bad events are likely to be (1) *permanent*, recurring over and over, (2) *pervasive*, occurring in many, as opposed to few, situations, and (3) *personal*, internally caused by one's own characteristics or behaviors. Optimists tend to take an opposite of these same bad events, treating them as *temporary*, *limited*, and *external*. Conducting study after study in a variety of settings with groups as diverse as elementary-school children and adult insurance salesmen, Seligman and his colleagues provided vast empirical support to Bruce Lee's observation that the confident athlete has "rationalized previous failures," effectively finding ways of "explaining" those failures in ways that protect one's confidence.

One of Seligman's classic studies took place at West Point during the summer of 1988 when he administered his Attributional Styles Questionnaire, an inventory designed to measure how optimistically or pessimistically one explains both positive and negative events, to the entire incoming class of new cadets at the start of their Cadet Basic Training, or CBT. Affectionately referred to in West Point slang as "Beast Barracks," or just "Beast," CBT is where twelve hundred young men and women, all of whom were outstanding achievers in high school, are subjected to six weeks of dawn-to-midnight military indoctrination. For those six weeks, through the hottest summer weather New York State has to offer, their emotional stability, their physical fitness, and their ability to quickly learn new skills are tested like never before. Seligman, who had long been studying how explanatory style influenced persistence under challenging conditions and eventual performance, could not have found a better setting. What did he find? Statistical tests conducted at the end of the cadets' plebe year showed that those cadets who quit the West Point program either during the summer CBT or during the following academic year (another period of high stress) had a significantly more pessimistic explanatory style than those new cadets who stuck it out and completed the year.

Seligman concluded in his original report on this study that "an individual with a pessimistic explanatory style is more likely to have problem-solving difficulties, and to become passive and give up when adversity is encountered." He also offered this hopeful note in the very last paragraph of his report: "explanatory style can be stably boosted by cognitive therapy. So it may be that preventative or remedial explanatory style training could help those cadets most at risk for helplessness against the inevitable setbacks that West Point life brings as well as the unique stresses involved in combat." While you may never experience the specific types of hardship and challenge faced by new cadets at Beast Barracks or during a plebe year at West Point, you may be tempted to explain to yourself the inevitable bad events that occur in your life by saying "It's me" (internal), "it's going to last forever" (permanent), and "it's going to undermine everything I do" (pervasive). Doing so puts you at a greater risk of not making it through your next trial because it destroys your confidence. But no matter what your tendency is right now, you are not doomed to a lifetime of pessimism. You can change the way you look at bad events and protect your mental bank account by treating those inevitable bad events as *temporary* ("it's just this one time"), *limited* ("it's just in this one place"), and *nonrepresentative* ("that's not the truth about me").

Safeguard #2: The Last Word—How to Win the Battle with Your Own Negative Thinkings

It often seems that if I had a dollar for every time a client asked me: "Doc, how do I stop all my negative thoughts?" I could happily retire on my own Caribbean island. No single question has been put to me more often, and no single question gets asked with more passion and more urgency. This comes as no surprise; the quest for relief from the constant chorus of negative voices that alternately whispers, whines, scolds, and screams at us has been undertaken by psychologists and philosophers for centuries. And while it's impossible to accurately quantify just how many thoughts you might have in a given day and what percentage of those thoughts must be countered to prevent them from undermining your confidence, you have probably experienced plenty of internally generated self-criticism, self-doubt, self-questioning, and negative self-labeling (*You are such a choke artist!*). These internally generated thoughts are every bit as lethal to your mental bank account as the external setbacks and negative events that the imperfect physical world we all live in might throw at you. Here's how to take some control over those voices and further protect your confidence. It's a process I call "Getting in the Last Word."

You've probably had the experience of getting into an argument or disagreement with a family member, or a coworker, or a teammate, or your supervisor at work. The topic or the disagreed-upon issue could have been about anything, but two things about that encounter were constant—there was a back-and-forth exchange between you and the other person, and whoever made the final statement, whoever got in the last word, usually "won" the argument. Dealing with your own internal negative thoughts, fears, and worries is very much the same—two competing opinions are vying for control of your mind and one of them is going to emerge as dominant. One voice is urging you forward, keeping you focused on what needs to be done, and building you up, while another is criticizing each step you take and distracting you with worries about bad future outcomes. Which voice will "win the moment"?—the one that speaks last and gets in the last word. To win those moments, follow these three steps.

Step One: Acknowledge It

As much as we'd all prefer not to deal with the way our own mind seems to attack itself ("I'm often my own worst enemy, Doc!"), the first step in defeating these internal enemies is to *notice them the moment they arrive.* Common sense tells us that you can't win a fight that you don't know you're in, so if you hope to defeat any internal enemy of doubt and fear, you must first acknowledge that enemy's presence. While some of my clients and trainees know exactly where and when their internal negativity tends to kick in (e.g., "the minute I walk in the locker room," "as soon as I miss two in a row on any shooting drill," "as soon as so-and-so enters my office"), others find the enemy more diffuse and elusive. Whatever the case is for you, the first rule is keep your internal radar up and alert so you detect the arrival of any negative thought quickly. Once you notice that voice chiming in, acknowledge its presence and bring it out of the background. For most of us, the voice of negativity prefers to hide in the shadows and pester us from a distance, but you can bring it out into the open by saying *Okay, I hear you.* Just doing that puts you in charge of the encounter; you're not just a victim listening to that voice as it talks, but now you are on the attack talking back to it.

Let's say you're Nick Vandam, a veteran professional triathlete competing in the 2012 Military World Triathlon Championship. As a former collegiate swimmer who transitioned to triathlon as a junior at West Point, you are proud of your ability in the water and feel that the swim portion of any triathlon is your strongest. You joined the Army's World Class Athlete Program after graduating in 2004, you've logged thousands of hours of quality training under quality coaches and completed hundreds of races.

But midway through the swim portion of this race, instead of continuing to surge through the water as you expect to do in this your strongest leg, you suddenly began to hyperventilate. A voice in your mind screams out, *Something's wrong! I'm missing breaths and getting beat when I should be way out in front! This is supposed to be my best event!* You stop dead in the water and begin to panic as other competitors flow past you on both sides. At this moment you have a choice—either abandon the race altogether or find a way to recover your confidence and get back in it. Now is the time for your mental training to kick in and talk back to that voice of fear and doubt. You begin by acknowledging that your confidence and your identity as an athlete is under attack (you're supposed to be a great swimmer, right?). *Stop playing these psychological games with yourself* is the thought you go to, recognizing the human tendency to sometimes be your own worst enemy, and refusing to let that voice continue unchallenged.

Step Two: Silence It

Now that you have identified the intruding thought and have it squarely in your sights, you can take the next step and effectively eliminate it. This is the same as saying, "No, you haven't got it right" when your obnoxious younger (or older) sibling says something irritating. It's the same as emphatically telling the family dog to "Stop it!" when she growls at a passing neighbor. What you are doing is further exerting your control over the conversation and neutralizing that mental thief who is trying to steal some of your confidence. Simply say to yourself *Stop!* in a determined internal tone of voice. Add the image of a stop sign or a police warning light. Or have a rubber band on your wrist to pull and release so that it snaps back onto your wrist reminding you to "Snap out of it!" My late friend Ken Ravizza taught his students to imagine a toilet being flushed, and he would even bring a tiny palm-size toilet toy, complete with lever and flushing noise, into the dugout of Major League Baseball teams so that any player could readily "flush" the memory of a bad at bat or an error in the field from his mind. Use whatever picture, symbol, or action works for you to communicate the end, the departure, or the destruction of that negative thought. That trigger, whatever it is, establishes a clean break from the unproductive, disabling thought and clears the way for more effective thinking.

Continuing with our triathlete example, this silencing step is dirt simple. You emphatically scream *STOP!* to yourself to take total command of your mind and the moment.

Step 3. Replace It

Time to decisively counterattack! Like a boxer who has read his opponent's attack and successfully deflected the incoming punches, or a trial lawyer who has refuted the evidence presented by her adversary, you can now step in with your own counterpunch or closing argument to get in the last word and win a small but significant First Victory. What will that counterpunch or closing argument be? How about one of the memories you've been deliberately depositing through your Top Ten, your daily E-S-P, or your regular IPR reflections? A quick glance at any of your journal pages will provide you with plenty of suitable "last words" ("I hit 8 of 10 three-point baskets vs. a tight defender"). How about those affirmation statements you've been repeating each time you walk through a doorway? Any one of them can serve as an effective counter to an attack on your mental bank account (e.g., *I make great decisions in difficult situations*). The worrisome thought of the struggling math student, *I'll never understand this stuff*, can be stopped and replaced immediately with *I've learned new formulas before and I can do it again*. The voice of doubt experienced by the soccer player who has misplayed her last two touches on the ball (*What's wrong with me today?*) can be replaced with *I'm okay. Just get after the next one*. With practice, each and every attack on your confidence that your mind unfortunately produces (yes, we are indeed our own worst enemies at times), can be acknowledged, silenced, and replaced. Even the ones that involve your own physical safety and survival.

Finishing our triathlon example, you replace all those fears and doubts with a simple mantra that you've practiced hundreds of time in both training situations and past competitions. It's nothing but a single deep breath and then the strong motivational directive to yourself, *Refocus and Go!* That's all you need to resume your stroke and work your way back into the race. Before you know it you're finishing the swim leg seventh out of the sixty-four contestants and only five seconds behind the leader, after having nearly quit. You finish the race as the top American, tenth overall, with your best ever time for that distance. As good as it feels to finish well and record a personal best time, there's a special sense of accomplishment with having won that personal mental battle against the voices of doubt and fear. "Everyone feels great until they get beat up a little," Nick Vandam told me years later, "and that's where the real competition starts."

Lieutenant Colonel Jonas Anazagasty couldn't agree more. "Confidence," he says, "isn't the absence of doubt, it's the way you respond to doubt." As he was participating in the Combat

Diver Qualification Course (CDQC) during his senior year as a West Point cadet, his confidence was attacked daily. The CDQC is a physically demanding and technically complicated Special Forces qualification course in underwater navigation and performance, where trainees spend five weeks learning to successfully operate scuba gear in the most demanding conditions. During the appropriately named One Man Confidence Test, a CDQC graduation requirement, trainees enter the water wearing full scuba gear but also a blacked-out mask. With their vision completely obstructed, they spend the next twenty minutes underwater, and between ten and fifteen times in those twenty minutes, they will be flipped over, knocked off balance, and have their scuba gear turned off, pulled apart, and otherwise continually compromised by instructors. All this is a test to see how well they handle unpredictable adversity. Between ten and fifteen times in those twenty minutes, they will have to hold their breath for thirty to sixty seconds while readjusting the breathing gear and regaining equilibrium.

As Jonas Anazagasty settled into the water for his second attempt at the One Man Confidence Test he knew he was in a must-win situation. He had failed his first attempt, and one more meant failing the entire CDQC course, but having practiced keeping failures in their proper perspective (Safeguard #1), and having practiced calmly acknowledging, silencing, and replacing negative thoughts with *Just stay calm, you're okay* even when his lungs were screaming for air, Anazagasty passed with flying colors. "This success," he says, "proved that I could deal with unexpected adversity down the line." His present position as commander of the US Army's Fourth Ranger Training Battalion brings plenty of that adversity, but Lieutenant Colonel

Anazagasty gets in the last word whenever it arrives with "I can handle this."

This simple and commonsense process of *Getting in the Last Word* to ensure that your self-talk remains constructive has been a staple of cognitive therapy going back to the 1970s, when pioneering psychologists Aaron Beck and Albert Ellis stepped away from the prevailing psychoanalytic- and behaviorism-based theories of the times. Beck and Ellis led the movement in psychology to propose that individuals' conscious thoughts, rather than their unconscious motives, were the primary source of their discomfort, and that these thoughts could be rationally examined for their effectiveness, disputed when they were determined to cause one difficulty, and then replaced with more empowering thoughts. This was a major break from the status quo of the times in that it put individuals more in charge of their own lives, and it set the stage for the emergence of a more positive orientation to psychology, later led by Martin Seligman.

By the late 1980s, research into the effectiveness of positive self-talk on performance had begun in earnest, and since then dozens of studies have documented its effectiveness in enhancing both the subjective levels of individual confidence and objective levels of performance in tasks as varied as dart throwing, skiing, distance running, endurance cycling, marksmanship, and basketball shooting. A comprehensive meta-analysis of thirty-seven studies on the effectiveness of self-talk, conducted in 2011 by a team at the University of Thessaly in Greece, concluded that "Overall, self-talk was confirmed to be an effective strategy for enhancing task performance in sport." Another meta-analysis completed the same year by a team from the University of Bangor in the United Kingdom concluded "the existent evidence base does suggest that self-talk has beneficial effects on cognition (in particular, concentration and focus-related variables), cognitive anxiety, and the technical execution of movement skills." Outside of the sport and performance skill arena, self-talk interventions have proven to be effective for public speaking, body image improvement, and stress management.

The most useful and relevant research finding on the value of self-talk that I know of, however, comes from the experience of journalist Alex Hutchinson, author of the 2018 book *Endure*, in which he explored (as stated in the book's subtitle) "the curiously elastic limits of human performance." Hutchinson graduated from McGill University in Montreal, Canada, in 1997, where he competed as a 1,500 m runner (he qualified for the Canadian Olympic Trials twice). During his time on the McGill team he listened to a sport psychologist suggest that he talk back to himself whenever he heard an inner voice bring up the fears and worries that are part of every middle-distance and long-distance runner's life—thoughts like *This pace is too fast* or *I wonder if I can keep up*—and replace that voice with one that keeps up a constant barrage of helpful advice like *Hang tough* and *Push through it*. But Hutchinson and his teammates thought all this "motivational self-talk" advice was nonsense and never practiced it. Like many athletes in their physical prime, they figured success in their sport was simply a matter of who had the biggest set of lungs and the best-trained muscles. The mental battle of confidence versus self-doubt was something they just didn't want to believe in at the time.

Fast-forward twenty years. Hutchinson has completed a PhD

in physics, run dozens of marathons and ultramarathons (races longer than thirty-five miles), and exhaustively researched every factor that contributes to endurance performance—training regimens, dietary practices, and yes—sport psychology. His investigations brought him into contact with Samuele Marcora at the University of Kent, whose research on human endurance had included studies of the effects of motivational self-talk, the same procedure Hutchinson and his college teammates had scoffed at, on time to exhaustion—the length of time an athlete can go all out pedaling a stationary bike before he calls it quits. In a 2014 study Marcora and his colleagues took twenty-four trained cyclists, determined their time to exhaustion, and then had half of them practice using positive phrases during both the beginning and ending phases of their training sessions over the next two weeks, essentially practicing how to get in the last word when boredom, fatigue, or pain knocked on the door of their mental bank account. When tested to exhaustion again two weeks later, this trained group lasted 18 percent longer than on their previous test, while the control group, who hadn't been practicing how to get in the last word, didn't change at all. Also, the positive self-talk group rated their perceived exertion (how hard they felt they were working) as lower throughout the test. Lasted 18 percent longer? And felt better while doing it? How significant is an 18 percent improvement in *anything*? Who wouldn't want an 18 percent improvement in their performance, whether we're talking batting average, shooting percentage, recovery time, or deals successfully completed? Seeing these data, and considering them in the context of everything else he had investigated, Hutchinson changed his mind about the value of talking back to that voice of worry and getting in that last word. On page 260 of *Endure*, only seven pages from the book's end, he writes, "If I could go back in time and alter the course of my own running career, after a decade of writing about the latest research in endurance training, the single biggest piece of advice I would give to my doubt-filled younger self would be to pursue motivational self-talk training with diligence and no snickering." I urge all my clients and students to take a hint from Hutchinson's experience and practice Getting in the Last Word regularly.

Unfortunately, not everyone does so. Years ago, a professional hockey player signed a dream contract worth tens of millions of dollars with an NHL team in a major market. He had all the physical talent one could ask for, and the team he signed with considered him to be the long-awaited answer to their prayers for an impact player at a key position. But it didn't work out. Almost immediately, the player found himself effectively paralyzed by the fear brought on by a constant stream of negative self-talk, and the quality of his play suffered accordingly. When I suggested to him that he talk back to those attacking voices, he cocked his head and looked at me as if I was absolutely crazy. The idea that he could exert control over his thoughts and use them to put himself in a better emotional state was something he had never considered and something that was utterly incomprehensible. For him, those voices were the boss and he was just a passive listener. Ultimately, his damaging mental habit of allowing his negative self-talk to continue unchallenged proved to be his undoing and brought his playing career to an early end. This is admittedly an extreme example, but if negative self-talk wasn't such a powerful confidence thief, I wouldn't have so many clients telling me at our very first meeting that "so much of the time I'm my own worst enemy."

Making It All Stick

As simple and direct as Getting in the Last Word is, three factors make its consistent application difficult. Having some awareness of these factors will make it easier for you to Acknowledge, Silence, and Replace your negative thoughts.

The first is the widespread misconception that "if you're truly confident you won't have any negative self-talk." We've been led to believe that confident and successful people, the champions like LeBron James or Michael Phelps, are magically endowed with a bulletproof mind that is simply immune to negative thinking, that they have permanently silenced their voices of fear and insecurity. By contrast, all the rest of us who are striving for success in our own respective fields readily admit that we have plenty of negative self-talk and because we have it, we conclude that we can't be as confident as those champions always seem to be. This is nonsense. Those "champions," in fact, have plenty of their own negative self-talk, and it often sounds off at the worst possible moments. Tennis legend Arthur Ashe acknowledged to me during a chance meeting in 1980 in New York that he once walked onto Center Court at Wimbledon for a semifinal match thinking *What if I don't get any of my first serves in today?* Golfing legend Bobby Jones is said to have stood over a putt no longer than three inches on the final hole of the 1926 US Open Championship where he was set to win and thinking *What if I stub my putter into the turf and fail to move the ball?* American wrestler John Peterson, who had previously won a silver medal at the 1972 Olympics, was walking into the arena at the 1976 Olympics when he caught a glimpse of himself on a closed-circuit TV and thought, *What if I get pinned with the whole world watching?* The only difference between these "champions" (who each won their respective competitions that day) and the rest of us is how they respond to those voices of fear and worry. The "average" person, like the hockey player mentioned above, allows those voices to whisper and scream continually. The "champions," on the other hand, hear those same voices, as often and as loud as everyone else, but for them, each utterance of those voices is a signal that it's time to tighten their minds down and replace the attacking thought or voice with a helpful one. Just as courage is not the absence of fear but the proper action in the presence of fear, confidence, as Lieutenant Colonel Anazagasty observed, is not the absence of self-doubt but the ongoing resistance to it, the proper thinking in its presence.

A second factor to understand about Getting in the Last Word is the unfortunate fact that you will have to do it over and over again, and continue doing it over and over again, until you retire from your profession, sport, or field of study. Just like the old arcade game Whac-A-Mole, those voices out to rob your confidence bank account are going to pop up again and again no matter how many times you smack them down. In this way, winning your First Victory of confidence is different from winning a decisive battle that ends the war once and for all (like the surrenders of Germany and Japan to end World War II). Unlike those external enemies in conventional warfare, the internal enemies of doubt, fear, and insecurity can never be ultimately defeated; they are just aspects of the human condition that no one ultimately escapes from. The lie that has been pushed to us through films, TV, and other media is that once some fairy godmother or wise mentor shares with us a secret to success, all our fears and insecurities vanish, letting us live happily ever after. This is a lie. Don't buy into it.

To help my clients rise above the universality and tenacity of self-doubt, I often break down a video clip featuring mixed martial arts fighter and coach Chael Sonnen counseling a younger fighter, Uriah Hall, during an episode of 2013 *The Ultimate Fighter* reality TV series. In this clip, Hall admits to his coach that he occasionally loses his confidence. "It's like a little drop of poison," Hall confesses, "that gets in your head and affects everything else." Coach Sonnen sympathetically replies that he experienced the same thing in his career and then shares two insights that he learned through work with a sport psychology pro. Number one: it wasn't just him—he used to think that he was somehow flawed because he experienced these doubts, but he came to realize that every fighter has them; they are just a normal part of the pursuit of success. Number two: they never go away entirely. Sonnen relates to Hall a conversation with Hall of Fame MMA fighter Randy Couture, in which Couture admitted that he could never defeat "the second-guessing and the negative voices in his head but that *he could compete with them*" (italics added). Sonnen congratulates his young fighter for acknowledging his ongoing mental battle and reminds him that he has a choice as to how he will respond to the inevitable invasion of self-doubt: "When doubt seeps in, you've got two roads . . . one road leads to victory—move your feet, keep your hands up, stay off the bottom, and the other road leads to failure." That's good advice—you'll never have that decisive end-of-the-war victory. But you can have many small victories hour after hour and day after day, as you acknowledge, stop, and replace each negative thought as it arises. These small victories are the ones that matter the most, so take pride in how steadily and consistently you win them. As long as you are standing up to your negative thoughts and replacing them, you are winning.

A final factor that can affect Getting in the Last Word is just how sneaky that enemy of your confidence can be. It seems to know exactly which single sport skill, professional competency, or personal relationship skill you are most insecure about, and it attacks you right there, where it's bound to hurt the most. When you see a coworker or teammate (or worse yet, a competitor) who seems to effortlessly execute that one skill or task that always seems to elude you, that enemy of your confidence will pick right up on it and attack you. The competitive swimmer who has been working hard on his flip turns can watch a rival power through a practice turn and be attacked by *I wish I could do it that well.* The graduate student who has never excelled in math is likely to be attacked by *The stats midterm is gonna be killer.* Just like the previous two factors, this one is just a part of being human. We are all intensely aware and rather sensitive about the personal shortcomings that influence our performance. But we can choose to recognize these shortcomings, get to work improving them, and while we are working on them, refuse to listen to the voice that suggests they are going to prevent us from succeeding.

There's a humorous scene in the 1994 baseball comedy movie *Major League II* where the team's owner, who actually wants her team to lose so she can more easily relocate it to another city, strides into the team's locker room moments before the team is to take the field for an important playoff game. Dressed in a glittering black evening gown, the owner (played by Margaret Whitton) walks over to the key players and reminds each of them of their worst individual baseball statistic. "No way you won't improve on the .138 you're hitting with runners in scoring position," she tells one player. "I sure you've put that one-for-eighteen performance from last year's playoffs far behind you," she tells another. As ridiculous as the scene is (no major sports team owner ever wants their team to lose), I always get an understanding nod when I show it to a client, because we each have our own internal "figure in black" who can show up at all the wrong times and remind us of something that we've failed at, done wrong, or underperformed in. This is nothing to be ashamed of or afraid of. The figure in black may even remind us of something that we can become better at. But indulging that figure in black and allowing it to remain and influence the running total of thoughts in our mental bank account is a sure way to undermine confidence and performance. Listen to that figure for just a moment, but if it doesn't provide you with something helpful, exercise your mental filter and *acknowledge* that you are under attack, *silence* the negative voice, and *replace* it with a thought that produces energy, optimism, and enthusiasm.

Safeguard #3: The "Shooter's Mentality": How to GAIN Confidence When Bad Things Happen

If you're a little dismayed with the responsibility of protecting your confidence from life's inevitable setbacks, the reality of human imperfection, and all those recurring attacks by your own negative thinking, I have an uplifting, empowering concept, a third safeguard for you as the conclusion to this chapter. The safeguards described so far can certainly prevent your precious mental bank account from being drained away by the setbacks you encounter and the negative thoughts those setbacks can create, but why not turn your mental filter up another notch so instead of just holding your account steady it actually grows in the presence of difficulties? How about using your mental filter to actually *gain confidence even while you are making mistakes, experiencing setbacks*, and *performing below your peak*? You can do this by combining the two safeguards mentioned above, adding a dash of highly selective amnesia, and then topping it off with some highly selective expectation. This results in what's called the *Shooter's Mentality*, the attitude that all the great "shooters" in any sport (basketball, hockey, soccer, lacrosse) and in any field (science, sales, entrepreneurship) consistently use to continually put themselves in a position to break through and win. The Shooter's Mentality consists of two habits of thought that might seem at first to be mutually exclusive, but which can indeed combine to bring about many First Victories. The first of the two habits is the tendency to think that any mistake or setback is actually bringing you closer to success rather than keeping you away from it. The second habit is the tendency to think that once any success is achieved it will continue and will make other successes possible. In the Shooter's Mentality, misses are indeed interpreted as temporary, limited, and nonrepresentative, but they are also seen as signals that a return to fortune is about to happen. Successes, on the other hand, are interpreted as being permanent ("It's going to happen again"), and as universal ("Now other good things are going to happen too"). If you are willing to cultivate these habits (and the will to do so is the only thing you need), you will have a thermonuclear psychological weapon on your side.

Stephen Curry, the scoring ace of the NBA's Golden State Warriors, certainly has it. His coach, Steve Kerr, says, "All Steph knows is 'I'm gonna shoot' and if he misses, he'll be all right because *he knows he'll make the next one*" (italics added). In Curry's highly selective and functionally constructive mind, *any missed shot just makes the next shot more likely to go in*, rather than bringing on a flurry of worry that he might be in for a frustrating night. That same highly selective and supportive mind also assumes that whenever he's "on" and making shot after shot, he'll stay that way all night, rather than wondering when his luck might run out. It's not necessarily logical to think that way, but it certainly is helpful, and Stephen Curry wouldn't have it any other way.

Thomas Edison certainly had it throughout his epic career of developing the incandescent light bulb, the electric storage battery, and other technologies we take for granted today. Rather than regard any unsuccessful trial or test as a "failure" (legend has it that he had over nine thousand "failures" before he got the battery to work), Edison had the attitude that every such test provided him with valuable information that could only help get him closer and closer to a final solution. Instead of draining away his enthusiasm and energy, each so-called failure only made him more certain that success was right around the corner. We can all be thankful for Edison's relentless optimism in the face of repeated "failures."

Tiger Woods also had it during his decade atop the professional golf world. During one tournament in this period, when he was twelve strokes behind the leader after the third of four rounds of play, Woods was asked by a reporter how he planned to prepare for his next tournament coming up the following week. Woods replied that he wasn't thinking about the next tournament at all, that he was putting all his energy into getting ready to play his final round tomorrow. "But you're twelve strokes back, you're out of it." the reporter pointed out. "That's not how I look at it," said Woods. "I know I have a 55 in me somewhere, and if I bring that out, and some of these guys on the leader board falter, I can still win this thing." For Woods, and other performers like him, the idea that a period of substandard play might continue doesn't find a place in his thinking. Instead, he (and the others like him) is firmly convinced that his very next effort will be a smashing success. Unlike the average individual, each of these great performers refuses to get caught up emotionally in their misses, setbacks, and "failures," and only sees before him an ever-growing range of opportunities.

I first came across this concept in my sport psychology PhD program, when my adviser, Dr. Bob Rotella, shared the story of an exercise he had conducted with his graduate students some years earlier. Always curious about how successful athletes thought, Rotella had assembled a small group of the University of Virginia's best intercollegiate athletes and asked them to each describe their most confident moment of their sports career to his graduate student group. As you might expect, the athletes told one story after another of dominant, successful play: touchdowns scored, personal records set, and rivals defeated. That is, until it was Stuart Anderson's turn.

Unlike the other athletes on the panel, Anderson, a football player at the University of Virginia at the time, told the graduate students about a game in which he performed rather poorly, a high school basketball playoff game where, right up to the final minute of play, he had only made one of fourteen shots from the floor. His team had played well enough, despite his poor shooting, to tie the score, so his coach called a time-out and huddled the team to set up the final play. Naturally, because Anderson was off his normal game that night, the coach diagrammed a play to give another player the final, potentially game-winning shot. As he was doing so, however, Anderson interrupted him, saying, "No, Coach. Give me the ball. I want the shot!" At first the coach refused, but Anderson then said something, with utter seriousness, that changed the coach's mind, "Give me the ball. I want the shot. *I'm due to get one in.*" That final sentence convinced the coach that Anderson was certain he could make the crucial shot, so he revised the play accordingly and sent the team out onto the court. With the final seconds ticking down, Anderson was fed the ball and he sunk a perfect shot to win the game. As a result of his two-for-fifteen performance he was carried off the court on the shoulders of his joyous teammates.

Hearing this story the graduate students immediately asked Anderson what made him think he could make that last shot, given how poorly he had been doing up to that point in the game. Anderson replied that he had been a 50 percent shooter from the floor throughout his career, so in his mind after missing a couple of shots, he figured that his odds of making the next shots were better than 50 percent. That comment raised the eyebrows of the students, all of whom were studying statistics and research methodology. Anderson's thinking went against everything they were learning about the logic behind the science of probability. But Anderson continued. "After missing four or five in a row, I figured my odds were way better than 50 percent, and by the end of the game, after missing I don't know how many, I figured it just had to go in." Paradoxically, Anderson became *more confident* with each missed shot. This initially perplexed the students, but some of them began to sense that maybe Anderson was on to something by thinking that way, even though it was illogical from a scientific standpoint. One of them then asked, "So you're telling us that you think your chances actually get better each time you miss. What about when you've made several shots in a row? Does that make you think that you're due to miss the next one, to bring you back to your average?" Anderson replied, "No! If I'm on a hot streak I just think I'm going to make everything I look at, so I keep shooting!" This comment made sense to the students—if things are going well for you, just enjoy the ride for as long as it lasts.

But then one student raised his hand and asked, "How can you have it both ways? How can you think that your odds are getting better when you're missing shots but also think that the odds are in your favor when you're making every shots?" Anderson's reply was simple: "I don't know. That's just how I think."

That's the Shooter's Mentality—misses just make hits more likely, while hits just make more hits more likely. No, it isn't logical at all, but it contributes to that all-important sense of certainty at the moment of truth, and that certainty will always give you the best chance of succeeding. The moral of the story is that your thoughts about yourself and your performance, all of which find their way into your mental bank account, need not be dictated by strict, everyday logic. Stuart Anderson, Thomas Edison, and Jason Mascia, the functionally delusional but very successful car salesman we met in the last chapter, all choose to filter and selectively interpret what happens on the court, in the lab, and on the showroom floor in ways that support continued effort and enthusiasm. You could say that they each create their own "reality," their own unique mental environment, one that might not always seem "logical" to others but that allows their talents and skills the fullest opportunities for expression. Logic might argue that Tiger Woods (or any other golfer) should forget about winning a tournament when he's twelve strokes behind the leader, and that Thomas Edison should forget about developing a storage device for this barely understood substance called electricity after several thousand failed tests. There is no logical justification for Jason Mascia to believe that each and every person who comes into his dealership will buy a car that very day. The sales data, the hard facts of his industry, would argue against him thinking that way. So why operate from this state of mind, from this personal "reality"? Because it's what always gives him the best possible chance of making that sale. The same is true for Woods, Edison, and everyone else: when you operate from that *sense of certainty*, your natural talent and your trained-in skills and your accumulated experience all combine to bring forward the best in you at that moment. Will doing that guarantee a victory each and every time? Certainly not. Will it prevent human imperfection from affecting your work? Certainly not. It will, however, help you perform well *between those inevitable occurrences of human imperfection*, and that will always give you the best chance of winning any victory.

Summary

It's *your* confidence. You worked hard to build up your mental bank account by managing your memories, telling yourself constructive stories, and envisioning your desired future. But that bank account is vulnerable. Your confidence is fragile and has to be protected. Your teammates, your coworkers, and, yes, you yourself will make mistakes despite your best efforts and intentions, simply because you are human. The ongoing stream of mental chatter that your highly developed brain produces will always contain some questioning, some second-guessing, and some self-criticism. But each of the attacks on your confidence presents you with a choice to either engage your safeguards or allow the world, human imperfection, or negative thinking to control your sense of certainty. You can choose to treat all those external setbacks as Temporary, Limited, and Nonrepresentative, and you can choose to Get in the Last Word when your own self-doubt arises. And you can choose to embrace your own version of the Shooter's Mentality to gain confidence even while making mistakes. As long as you are engaging these safeguards, no matter how often you have to do so, you are winning the ongoing mental battle for your confidence. Each engagement of these safeguards, each application, is a First Victory won, a moment where you can be proud of knowing that you did what you could, that you controlled what you could control. And if history is any guide, your engagement of these safeguards will lead to some other victories as well.

CHAPTER SIX

Deciding to Be Different

Letting Your Bank Account Earn Maximum Interest

Pause and consider all that has been covered in the preceding chapters . . .

The nature of confidence and the inescapable fact that the ultimate source of confidence is the way you think about yourself, your life, and the things that happen to you in your life.

The need to be selective in your thinking, and the process for doing so—using your human free will to mentally filter in and emphasize the thoughts and memories that build energy, optimism, and enthusiasm while deliberately releasing or restructuring the thoughts and memories that do anything else.

The specific tools and techniques for building up your mental bank account: managing your memories, repeating the constructive stories, and envisioning your desired future.

The mental tools and techniques to prevent life's inevitable setbacks, human imperfections, and our own negative thoughts from drawing our mental bank account down: rationalizing the setbacks as temporary, limited, and nonrepresentative, and using each negative thought as a trigger to stop, cope, and take control.

All this is designed to help you win the First Victory of decisiveness and trust at the moment of truth when it's time to perform. By following these guidelines you will be less likely to fall back into analysis, judgment, and worry during any performance because you have (1) built up that large bank account of reasons to believe in yourself, (2) reduced the self-criticism and self-distraction to a minimum, and (3) kept your mind focused on what you want to accomplish and how you want to be rather than on what you fear or wish to avoid. These practices are effective and are well supported by science. Despite their effectiveness, however, they have yet to become mainstream teachings in schools and have yet to be generally accepted throughout modern society. As mentioned in the introduction, modern society is very ambivalent when it comes to confidence—it's personally important to have some, but socially deadly to have too much, and just how much is "just right" is never made clear.

In this chapter I challenge you to look carefully at some of the assumptions and ideas that are prevalent in our culture regarding the development and expression of confidence. It's likely that you were exposed to some ideas and understandings during your upbringing and early training that encouraged you to conform and fit in rather than stand out and express yourself fully. The process of "socialization," defined by the *Oxford Dictionary of English* as "the process of learning to behave in a way that is acceptable to society," is a double-edged sword. It provides safety and security, but it doesn't always encourage the search for your personal excellence. If you desire to find out how great you can be at your chosen sport, art, or profession and win a consistent First Victory, you are going to have to confront this negative aspect of socialization and decide to think *differently.* This chapter shows you how.

A Case Study in "Thinking Differently": Deion Sanders

To take our understanding and practice of the First Victory further, let's conduct a "case study" and examine the thinking process of a very confident and high-performing individual. I've always been a fan of case study research—going directly to the people who are exhibiting an interesting or valuable trait, studying them carefully, and then sharing the knowledge collected. One of my top ten recommended sport psychology books is *The Pursuit of Sporting Excellence: A Study of Sport's Highest Achievers,* written by British track athlete David Hemery (gold medalist, 1968 Olympics, 400 m hurdles). Why do I recommend it? Because it's about real people who distinguished themselves in the real world of human performance. Hemery interviewed fifty champions and top performers across a wide range of sports to determine "if there were common factors among sport's highest achievers which could be applied to anyone who aspired to fulfill their talent." Hemery's case studies provide valuable insights into what really works in the real world (and not surprisingly, 87 percent of the athletes interviewed said they had a very high degree of self-confidence). Some of the most influential business books have used a similar case study approach. In 1982, management consultants Tom Peters and Bob Waterman published *In Search of Excellence: Lessons from America's Best-Run Companies*, where they identified the characteristics and practices that separated a few highly successful companies from the majority of American businesses. Over the next fifteen years the book sold 4.5 million copies and remains to this day one of the classic works in the field. In 2001, Jim Collins published *Good to Great: Why Some Companies Make the Leap . . . and Others Don't.* This book also described the defining characteristics and practices of companies that separated them from the pack, and it too became a bestseller. What all three of these books had in common was their "case study" approach: rather than proceed from an idea or a theory about success and seek out examples as proof, these authors went directly to the successful companies and individuals and learned what they did that separated them from the others, what made them *different*.

In this spirit let's examine how a very successful individual built and maintained an impressive mental bank account to separate himself from his competition, and pay attention to just how *different* his thinking is from the mainstream. This examination will lead us to important questions about some of the underlying assumptions and beliefs that influence our mental bank account. Have those beliefs kept us from winning our First Victory by encouraging overanalysis and other forms of ineffective thinking, or have they helped us move forward and separate ourselves from competitors and opponents? Can we embrace new and more constructive beliefs that will help us keep winning our First Victory? Absolutely.

Meet our case study subject: Deion Sanders. Anyone who's played in the Super Bowl will tell you it's an incredible, never-to-be-forgotten experience. Anyone who's played in a World Series game will say the same. But there's only one person on earth who has played in BOTH the Super Bowl AND a World Series game. That person is Deion Sanders, the Hall of Fame football defensive back and former Major League Baseball outfielder. With two Super Bowl rings, eight Pro Bowl appearances, and, until 2014, an unbroken NFL record of nineteen returns for touchdowns, Sanders made good on his stated intention "I never wanted to be mediocre at anything, I wanted to be the absolute best." I'll leave it to football pundits to decide whether Sanders was the absolute best, but no one will ever argue that Sanders was even close to being mediocre. That word *mediocre* comes from the Latin *mediocris*, which means "middling, ordinary, unremarkable," terms that could never be applied to the man known as Prime Time and Neon Deion.

Sanders began his professional football career in 1989, playing for the Atlanta Falcons while also playing baseball for the Atlanta Braves. He played the 1994 NFL season with the San Francisco 49ers and helped them become Super Bowl Champion that year. It was during this season that he gave a remarkable (in my opinion) television interview on ESPN. In the course of a five-minute interview with former NFL quarterback Joe Theismann, Sanders made five statements that serve as beautiful examples for anyone seeking to win their First Victory.

The interview begins, after some sparkling highlight footage of Sanders intercepting passes and celebrating touchdowns, with him speaking plainly and simply, "I believe that I'm better than you," not with any bravado or swagger but with a calm matter-of-factness. Thirty seconds later, Sanders can be heard saying, "I will outplay you on that specific play," again as a plain-as-day statement. Barely twenty seconds later, Theismann the interviewer starts to set up his next question by beginning with "When the ball is up in the air . . ." but Sanders doesn't even let him finish. He interrupts Theismann and delivers one of the most insightful comments I've ever heard from any performer. Sanders says, referring to the ball in the air, "It's mine! It's for me! It's not for him, it's for me! It was meant to come to me when the ball is up in the air. I think that's the attitude a defensive back must have." Fourth, Sanders replies to a question about his foot technique at the line of scrimmage by saying, "I change it up all the time. I have them thinking about me. I change the whole philosophy in their mind. Now they have to worry about me and what I am going to do." Last, Sanders responds to a question about his new team, the 49ers, by saying, "They have to believe in themselves, and that's what I do, I help the secondary believe in themselves." Each of these comments provides a window into Sanders's mental filter and how he wins his First Victories. Examine each one carefully.

1. "I believe that I'm better than you"

Whether this statement is factually true or not doesn't really matter. Was Sanders indeed "better" than the wide receivers that he was assigned to cover? Again, I'll leave that up to the pundits and statisticians. What does matter, however, is that Sanders genuinely believes he's better. When he takes the field to play, he believes that he is better than his opponent of the moment, and that conviction frees him to execute with the absolute bare minimum of self-consciousness and the momentary hesitation that such mental chatter might produce. Is he "justified" in having that belief? Again, that's beside the point. Just as we have seen with the Shooter's Mentality, there is a time and a place for using logic to justify your inner feelings and beliefs, but there's also a time to throw logic aside and believe that the odds are always in your favor. I urge every client and student of mine to embrace and practice this habit of belief. Believing you are better than any opponent you face is a prerequisite for bringing out your best performance, whether that "opponent" is another human being, as in any athletic contest, or whether the "opponent" is a report to be written, a concerto to be played, or a blocked artery to be surgically bypassed. Are you willing to choose to believe that you are better than the opponent in that moment or will you choose to believe something else? First Victory Takeaway: Approach every task with the conviction that you will absolutely succeed.

2. *"I will outplay you on that specific play"*

When you play cornerback in the NFL, the rules are stacked against you. Wide receivers with world-class speed are coming at you full tilt along a preplanned route while you have to backpedal and then make an informed guess as to where exactly they are going. While remarkably gifted with natural speed, and having diligently trained to backpedal quickly, every cornerback,

Sanders included, is going to lose that footrace a few times every game. He's going to get beat on some plays, passes are going to be completed against him, and even the occasional touchdown will be scored by the man he was assigned to cover. But when this happens, Sanders practices Bruce Lee's rule of "rationalizing the failure" by treating it as *temporary* (it was just that one play) and *limited* (it was just in that one place). With that kind of selective memory, Sanders can forget if you happened to beat him on the last play and bring all his energy and talent to bear so he can beat you on *this specific play* right now. First Victory Takeaway: The only moment that matters is this one; live it free of any resistance from your past.

3. "It's mine! It's for me! It's not for him, it's for me. It was meant to come to me when the ball is up in the air. I think that's the attitude a defensive back must have."

At first glance this comment is utterly ridiculous; the ball in the air was most certainly not meant to come to Deion Sanders. The opposing quarterback, receivers, and coaching staff all spent hours of their weekly preparation designing and practicing plays that would keep the ball *away* from Sanders. But in his highly selective and functionally delusional mind, Sanders considers that ball in the air to be his and his alone.

I urge all my clients and students to think about that "ball in the air" as a situation with an as-of-yet undetermined outcome. When the football is up in the air spiraling along no one really knows for sure where it will land or what will ultimately happen to it. It might be caught, it might be dropped, it might be knocked out of the air, it might fall harmlessly to the turf with no one touching it. That uncertain situation can be viewed in a number of ways. It can be viewed somewhat neutrally, with a "We'll see what happens to it" attitude. It can also be viewed with a sense of dread and foreboding, "Uh-oh, this could be bad for me/us." But the uncertain situation of the ball in the air and indeed *any uncertain situation* can also be viewed through a functionally optimistic and confident mental filter as "This was meant to turn out in my/our favor."

Please pause for a moment and consider what "uncertain situations" you face every day in your work, or your sport practice, or your professional life. Do you share Deion Sanders's "It was meant to come to me" attitude about each of them? I counsel every athlete trying out for a team to think, *That roster spot was meant for me!* I counsel every athlete on a team to think, *That spot on the starting lineup in MINE!* And I counsel every starter on any team to think, *That all-league or all-American award is MINE!* Deion Sanders says, "I think that's the attitude a defensive back must have." Dr. Zinsser says, "I think that's the attitude every athlete, every professional, and every performer must have." First Victory Takeaway: Every uncertain situation is meant to come out in your favor, so act with absolute certainty.

4. "I change it up all the time. I have them thinking about me. I change the whole philosophy in their mind. Now they have to worry about me and what I am going to do."

I'm not sure if Sanders actually changes any of his opponents' "philosophy," but I agree completely with the principle he's working from. Rather than think, "Okay, I've gotta cover ________ (insert Randy Moss, Jerry Rice, Andre Rison, or any other top level wide receiver of Sanders's era), and make sure he doesn't burn us for big plays," Sanders's attitude come game time is "That Randy Moss (or whoever) is gonna have his hands full trying to beat me. He's gonna have to up his game a ton if he's gonna get anything on me." Rather than think that he has to elevate his game to match any opponent or succeed in any new situation, Sanders (and any mentally tough pro) takes the view that the new opponent or new situation has to come up to *his* present level of competence. The pressure is no longer on him; now it's on the other guy. Mind you, Sanders will put in plenty of preparation and study to be ready to perform at a high level, but he will also, and this is the key, *decide that his level of preparation is always enough, and that the pressure is now on the other guy.*

Who are your equivalents of the top-level wide receivers that Deion Sanders must cover? Who are your opponents and competitors, and how do you tend to think about them? Are they obstacles or forces that require you to "dig down deep" to find some special reserve of strength or smarts that will allow success? Do you put pressure on yourself by thinking *I have to really be great in order to beat so-and-so or make that deadline?* or do you allow yourself a little peace of mind by thinking *That opponent has to beat* me. *All I have to do is play my game/do my work and I'll be fine.* Is the pressure on you or is it on the other guy to beat you? Is the pressure on you to make that deadline or is the pressure on the deadline to try and crack your composure?

For an alternate look at this attitude, there is a wonderful scene from the movie *Men of Honor,* the true story of Carl Brashear, the first African American and the first amputee to earn the US Navy rank of master chief petty officer. Brashear suffered a severe leg injury midway through his thirty-year Navy career during the recovery of a nuclear warhead from the ocean floor off the coast of Spain. Rather than leave the navy with his dream of making master chief unfulfilled, he insisted that his injured leg be amputated below the knee, convinced that he could still be effective as a salvage diver once he trained himself to run, swim, and dive with the proper prosthetic. At the scene depicting his official hearing to determine his readiness to return to duty after surgery and rehab, Brashear's character, played by Cuba Gooding Jr., responds to a question from the presiding officer, a skeptical navy captain who doubts Brashear's fitness for the physically demanding job of a navy salvage diver. "You're almost forty, and you have one good leg," the officer starts. "Do you really think you can keep up with healthy divers half your age?" Brashear's answer, whether the invention of a Hollywood script writer or an actual statement at the time, comes right out of the Deion Sanders playbook: "The question is, sir, can *they* keep up with *me*!" First Victory Takeaway: The pressure's always on the other guy; just play your game the way you know how.

5. "They have to believe in themselves, and that's what I do. I help the secondary believe in themselves."

I have to be careful here. While Sanders exemplifies the self-confident athlete, I don't know what kind of teammate he was. While doing the research for this book I was unable to find comments from his former teammates complimenting him for his positive influence in the locker room or the meeting room. On the contrary, there are plenty of comments about how Sanders was far more concerned about his own stats and accolades than he was about the success of the teams he played on. So what are we to make of his statement "I help the secondary believe in themselves"? I think the First Victory Takeaway here is this: projecting an aura of genuine confidence (which Sanders most certainly did) and modeling optimism and mental resiliency whenever the breaks go against him or his team (which he also did) are both valuable contributions to any team or work group. We're all on one team or another (and in some cases, several). Even the individual sport athlete (golfer, tennis player) has a coach/mentor and most likely a partner or spouse. So does the individual artist and sole proprietor of every small business. Every musician, even those who perform solo, relies on a team of light, sound, and recording pros to make performances happen. Every surgeon works with team that includes an anesthesiologist and several nurses. These "teammates" who are working with you and supporting you should get the benefit of your belief in yourself. They don't deserve to be the ones who see you at your worst or whom you complain to in your low moments (save all that for your sport or performance psychologist). What do your teammates get from you? Would they say that you maintain a great attitude regardless of whether you or the team are riding a hot streak or struggling through a downturn? There's a very trite saying on this but it is absolutely true, as anyone who's ever been part of a team will agree: "Attitudes are contagious. Is yours worth catching?" Although I cannot say for certain, I'm pretty sure that Sanders's teammates in the defensive secondary fed off his confidence and played better because of it. First Victory Takeaway: Project an aura of genuine confidence, modeling optimism and mental resilience.

The Big First Victory Takeaway: Deciding to be Different

"Thanks for getting me back on track, Doc Z." The speaker was Danny Brière, a seventeen-year veteran of the National Hockey League, whose enduring legacy was his ability to play his best hockey at the biggest moments in the biggest games. At five foot eight and 170 pounds, Brière relied on speed, guile, and yes, confidence, to score 116 points in 124 Stanley Cup playoff games. Brière had been exposed to sport psychology early in his pro career as a member of the Phoenix Coyotes, and the confident mentality he developed helped him become a top-tier NHL player and captain of the Buffalo Sabres from 2005 to 2007. But when he signed with the Philadelphia Flyers in 2007, things changed. It was a new team in a new city with a new coach and a very big contract, and all those adjustments turned into distractions that brought Danny Brière's confidence down. Instead of taking the ice with the attitude that he was "twice the size of any other player," Brière began to doubt himself, and those doubts affected his performance and production.

But those doubts didn't last very long. Danny Brière and I went to work on his attitude and rebuilt his mental bank account with a steady diet of quality envisioning, constructive self-talk (e.g., *I deliver when the game's on the line*), and by adopting Deion Sanders's idea that "the ball was meant to come to me." In the final game of the 2010 regular season, which the Flyers had to win to make the playoffs, Danny Brière's confidence, the result of this unconventional but constructive thinking, came through. That game ended in a tie and proceeded to a shoot-out, where each team selects three players to take an unopposed penalty shot against the other team's goalie; the teams alternate shooters and the highest-scoring team wins the game. Danny Brière was the first shooter up for the Flyers, and as he prepared to charge down the ice and take on Henrik Lundqvist of the New York Rangers, the best goalie in the game, the magnitude of the moment crept up on him. "I knew this was a pretty big deal. I knew everyone was watching me, that our playoffs hopes were on the line, that there was a lot riding on what I was about to do." And then he changed his mind. "This is where I'm supposed to be . . . I was meant to be the difference maker . . . this moment was meant for me!" Brière took the puck, flashed it past Lundqvist, and the Flyers won. Brière went on to have one of the greatest post-season performances in hockey history—recording twelve goals and eighteen assists as the Flyers advanced to the 2010 Stanley Cup Finals.

Reflecting on that performance Danny Brière told me, "I can't thank you enough. Our work really made a difference, and it's weird that so few players think this way."

Brière is dead-on. Cultivating this inner sense of certainty and bringing it out when the game is on the line *works*. It's not a 100 percent guarantee of success at every turn, but it always gives you your best chance. So why do relatively few players (and relatively few people) fully embrace it and use it consistently?

Simple but challenging answer: because it goes against what is taught in schools and by society in general about how one should behave and how one should pursue success. Most people were brought up to share society's general distaste of outspoken and confident people like Muhammad Ali and were taught to think of them as too arrogant, too conceited, too full of themselves. Did you have a teacher in middle school, high school, or college who encouraged you to think that you were the best at what you do, that no matter what just happened you were going to succeed at your next opportunity, and that every uncertain situation was likely to come out in your favor as long as you put in the work? If you did, consider yourself very lucky.

Most people weren't so lucky. Most of us were "socialized" by well-meaning teachers, coaches, and other authority figures who encouraged us to find our place in the world and comfortably fit in rather than build the self-confidence that would help us stand out. Ask yourself these questions:

What were the key sources of your thinking habits, emotional tendencies, and beliefs about yourself? How did you learn to think the way you do?

Were the people and institutions who taught you and guided you during your formative years interested in helping you discover how (dare I say) awesome you could be, or were they more interested in making sure that you fit comfortably into the world of normalcy and didn't rock the boat too much?

What were you encouraged, in both explicit ways and in more subtle and implicit ways, to pursue—the full expression of your unique talents, or the safety and security that modern life, with all its technologies and conveniences, can provide?

What were you taught about *how to think* in order to pursue success and satisfaction?

I'd like to challenge you a little here and ask you to look carefully at some beliefs that clients and students of mine have brought into my office over the last thirty years, beliefs that they have acquired through socialization during their school years, early playing days, and formative training. I will describe how they encourage the kind of thinking that brings your mental bank account down instead of building it up. While no single one of these beliefs will break your confidence, their combined influence, reinforced over a period of years and often communicated by figures of authority, can prevent you from developing the effective thinking habits described in the previous chapters. Refusing to buy into these beliefs, resisting these articles of socialization privately within the confines of your own mind, will help you win one First Victory after another.

Limiting Belief #1: Remember Your Failures and Mistakes; Those Memories Motivate You to Improve

This belief puts a constant negative pull on your mental bank account, dropping the wrong memories into your running total of thoughts and effectively running your bank account down. You might feel a momentary surge of energy produced by the anger aroused by the memory of a past mistake, but that energy surge is short-lived and leaves a nasty residue that you have to work through to improve. There's a cleaner-burning and longer-lasting fuel source you can use—the memories of effort, success, and progress and the vision of the future you really want.

Limiting Belief #2: Always Be Your Own Harshest Critic

There are certainly times when you have to look yourself squarely in the eye and acknowledge your mistakes and shortcomings, but there are also times when you should do the opposite. Sadly, however, the myth persists that if a certain amount of seriousness and self-criticism is good for you, then a lot of it must be better. As a result, a steady stream of self-criticism can become your default setting, with you believing that you are actually helping yourself by maintaining it. In reality, though, you're just buying into mediocrity by beating yourself up and drawing your mental bank account down. Face it, if you criticize yourself all the time in practice and think about all the things you have to improve when you're away from practice, then you shouldn't be surprised when you don't feel any confidence on game day. Save the self-judgment and the self-criticism for the right moments when you are away from the action, the times when you can calmly and rationally acknowledge your weaknesses without condemning or belittling yourself.

Limiting Belief #3: Always Be Logical and Think Carefully About What You're Doing

School was basically an exercise in logical reasoning. The hours we spent learning the multiplication tables and the rules of grammar drilled into us the idea that everything could be broken down into its components and then logically reassembled. Maybe music and art class might have given us an opportunity to be spontaneous and creative, but often even these activities were taught as a set of rules and logical structures to follow. When it comes to confidence, however, logic isn't always helpful. Logic dictates that the best predictor of future behavior is past behavior, that the team or opponent that beat you last time will do it again, that the job task that has always been troublesome will continue to be so. If strict logic was indeed the answer to everything and had always been followed, the Wright brothers would have never gotten off the ground, and Roger Bannister would never have broken four minutes in the mile run. Being "logical" closes us off from creativity, joy, and the discovery of anything new, precisely those things that give our lives the greatest meaning.

Limiting Belief #4: Always Look for More Knowledge, More Information, More Practice Opportunities

Constantly looking for the latest and greatest technical tips and inside information pointers sure sounds like a great way to improve yourself, but there are a few downsides to it that aren't well understood. First, the belief that there is some "solution" out there to be discovered encourages you to look outside of yourself rather than look inward into the nature of your thinking habits. Is the limiting factor in your professional or performance life really a lack of knowledge (not knowing what to do or how to do it), or is it a lack of trust in yourself to do all the things you already know how to do? Maybe the best way to take your sport, passion, or profession to the next level is by looking inside instead. Second, this belief encourages you to become more preoccupied with mechanics and technical instruction, and as we've discussed throughout this book, too much of that can lead to overthinking and confusion. Taken a couple steps further, this belief can trap you into destructive perfectionism where you feel you're never enough no matter how much you know or how hard you've worked. The truth is that the fewer things you have in your mind when you perform, the more efficiently your brain and nervous system will operate.

Limiting Belief #5: You Better Be Really Good at Something Before You Become Confident About It

This belief plays a nasty trick on you. It prevents you from feeling certain about yourself by always kicking your confidence down the road, always putting it just beyond your reach. It makes you ask yourself dangerous questions such as "Am I good enough at _____?" "Have I done enough?" "What else could I have done to train/improve/prepare?" These questions open the door just wide enough for self-doubt to come crashing through. They prevent you from believing in yourself *now*, and you may not have the luxury of more time and the necessary training resources that your self-doubt keeps insisting you need. Besides the fact that you may need to believe in yourself *now*, the forever questioning brought on by this belief almost guarantees that you will *never* feel truly confident. It will always elude you because there is *always more that you could have done*. An important fact to realize here is that your state of mind, your sense of certainty, is a choice that you make regardless of the situation you are in, and regardless of the level of preparation you have achieved.

Limiting Belief #6: Worship the Gods: The Experts Know What's Best and the Winners Have Earned Your Reverence

I don't remember where I first heard it, but I've always gotten a kick out of the saying "If you have a hero, the best you can ever be is second place." We read of Washington, Lincoln, and other iconic figures throughout our childhood while social media and mass media advertising bombard us with a constant stream of images and messages designed to make us yearn to be like some movie star, billionaire, or sports hero. Perhaps some people are inspired to follow these examples, but many more people end up questioning themselves and doubting that they'll ever live up to such standards. Unfortunately, the images of these godlike heroes put out by print, broadcast, and now social media are generally false; they are created not to tell the truth about anyone but to sustain the media outlets through the sale of advertising. Why buy in to the image of an opponent or a competitor as it's portrayed in a magazine or a TV spot? Looking up to some icon on a pedestal teaches us to compare our records and accomplishments with those of our competitors and opponents. Result? We end up overestimating our competition and underestimating ourselves when we don't really need to.

Limiting Belief #7: Above All Else, Thou Shalt Not Screw Up. AKA: The Team That Makes the Fewest Mistakes Wins

I have found that nothing erodes and destroys confidence more than the fear of making mistakes in performance. In a truly ironic way, we are all socialized to fear making mistakes so much that we tie ourselves up in knots trying to avoid them and end up doing worse. When the fear of making a mistake is in your mind, you become cautious instead of assertive, reserved instead of intense, overly analytic instead of natural and flowing. This belief implies that great players and performers almost never make errors, and if you buy into that you are caught in another vicious mental trap: once you make one or two errors as you perform, then you're automatically out of it; your shot at greatness is over. Since those mistakes are inevitable, human imperfection being what it is, believing that "the team that makes the fewest mistakes wins" puts you in a steady state of tension and worry.

Each of the preceding confidence-undermining elements of socialization can be resisted. These limiting beliefs only have power if you choose to endorse them. You can choose some alternative beliefs about the pursuit of success and satisfaction that may be different from those that you grew up with but will help you develop the sense of certainty that always gives you the best chance of success. You can choose to think *differently than most people* and when you do so, you give yourself a chance to *perform differently than most people.* In that spirit, here are seven alternatives to the limiting beliefs described above, beliefs to adopt and embrace to help you develop your own Deion Sanders/Danny Brière level of confidence. Think of them as expanded contribution limits to your mental bank account and special interest rates that allow your account to grow at the maximum rate possible.

First Victory Alternative #1: Remember What You Want More Of, That Alters Your Brain and Body So You'll Get More of It

Be honest with yourself. What creates more eagerness and motivation to move forward, mentally replaying that narrow loss you suffered and feeling its sting, or mentally replaying one of your most successful moments? A cadet on the West Point wrestling team replied this way: "When I remember my losses, I feel this huge sensation of fatigue, but when I remember my wins, I feel this lightness and energy. It's interesting because I know I was just as physically drained after the wins as I was after the losses, but those memories feel entirely different." This statement is supported by the latest science. A significant body of research has demonstrated that "recalling autobiographical memories that have a positive content," or in plainer language simply "remembering the good times" reduces the activity of the brain circuits that create the sensation of stress (the hypothalamic–pituitary–adrenal axis). Other studies indicate that recalling positive memories make you less susceptible to depression. Neither stress nor depression will help you learn new skills, improve your present skills, or perform under pressure. On the other hand, joy and excitement, the opposites of stress and depression, have been shown to facilitate learning and performance. So as the old rock 'n' roll song goes, "Let the good times roll!" Cultivate the habit of dropping into your brain ideas about and images of what you want more of.

First Victory Alternative Belief #2: Always Be Your Own Best (and Most Honest) Friend

You always stick up for your best friend, right? You accept his/her imperfections and weaknesses and still bring the love. You encourage and support him/her no matter what. Even if that person really messes up something, you stay on their side. During the times where something goes really badly, you're likely to take your best friend aside and say something like, "Look, I know you didn't mean to, but you really messed that up and you're going to have to fix it. It might be tough, but I know you can do it. Let me know how I can help." Usually after a talk like this, your best friend heaves a sigh and gets to work, knowing that you're still in their corner. They are really glad to have a friend like you.

But do you treat *yourself* that same way? Do you give yourself that same degree of support, that same level of what is called in psychology talk "unconditional positive regard"? I've found that very few people do; they have plenty of compassion for their best friends, especially when those friends are suffering, but they have little compassion for themselves in their own moments of pain and difficulty. Why? Because they've accepted the lie that accepting their own imperfections, that turning their compassion inward, will make them complacent, lazy, and lead to poor performances.

If you've accepted this lie in your life, take a tip from Helen Maroulis, who became the first American woman to win gold in Olympic wrestling at the 2016 Rio Olympic Games, and she did it by defeating Japan's Saori Yoshida, the most successful female wrestler in history (who hadn't lost a match in international competition in the previous sixteen years!). With four weeks to go before the Olympic Trials in 2016, Maroulis still had fifteen pounds to lose in order to make her required 53 kg weight, and the USA training team was concerned. While still training hard and dieting relentlessly, Maroulis found ways to be her own best friend and take some pressure off herself by spending time with her boyfriend, taking a few trips to the beach, and conducting youth wrestling clinics during these critical weeks.

As she explained in her History Channel documentary, "You have to step back and be relaxed and balanced. It's always about scaling things back and getting perspective. . . . Whether I do or don't get it (the spot on the Olympic team), it doesn't affect who I am." Then, at the 2016 Olympic Trials, Maroulis truly became her own best friend by finally forgiving herself for not making the previous Olympic Team in 2012, "and that was the last thing that had to get hashed out before I could just go out there and compete." And compete she did, defeating five opponents by a combined score of 64–2 and setting herself up for success at the Games. When I interviewed Helen in 2018 before an audience of West Point cadet athletes and their coaches, she further emphasized the importance of being your own best friend and accepting your weaknesses just as you would those of your friends.

"I went back to my journal five days right before the Olympics and I realized, wow, I'm looking for perfection, and I'm *never* going to find it. It's not a bad thing to look for perfection, but what's bad is the attitude that I had, meaning that I was never pleased with what I had. I asked myself, What am I going to do with perfection? What do I really want out of perfection? What I want out of perfection is excellence. Can I still have excellence without finding perfection? . . . Yes. All I needed to do was be *enough* to outscore my opponent, to be enough to win that match. And I say this all the time—I won with all of my strengths, and I won with all of my weaknesses. And I think sometimes we maybe lose sight of that. That we can't win, or we can't achieve things with those weaknesses. But you're always going to achieve things with those weaknesses."

Acceptance. Forgiveness. Compassion. These valuable methods

of supporting your best friend's pursuit of excellence work for your own pursuit as well. Use them.

First Victory Alternative Belief #3: Use Both Logic and Creative Fantasy to Create Your Own Reality

Just as there is a time and place for self-criticism, there is a time and place for careful, logical analysis. There is also a time and place, as discussed above, for self-compassion, and most definitively, a time and a place to throw realism and logic out the window and trust whatever skills and abilities you have. I urge my clients to limit their careful, logical thinking to only those practice drills and practice activities that involve a single skill, like simple passing and catching in basketball or simple "pads low, come off the ball fast" drills in football. The rest of the practice, when performing complex drills involving many skills simultaneously, I urge them to turn off their careful analytical minds and "look and do . . . sense and react." Why? Because the mental process of conscious analysis interferes with the smooth execution of skilled movements (sports, performing arts, surgery) and the automatic retrieval of remembered information (testing, taking questions from an audience, making a counterargument). The human mind's conscious analytic capacity is indeed wonderful and valuable, but just as wonderful and equally valuable is the mind's unconscious competence.

Sadly, most people only develop their logical minds, missing out on the possibilities that they might have experienced had they only allowed themselves to be a little more playful, a little more creative, a little less "realistic." Most people will never attempt anything unless their logical analysis of the situation tells them that they have at least a 50/50 chance of succeeding. Donna McAleer, West Point class of 1987, is not one of those "most people." She left her corporate job and spent two years training full-time to make the Olympic bobsled team despite never having been a power athlete. She twice ran for the US Congress as a Democrat in Utah, a state that ranks forty-third in the number of women in elected offices. She knew that neither of these goals were "realistic," but she pursued them with conviction and energy, firmly believing that if she worked both hard and smart, she'd come out on top in the end. Skeptics will say that since she didn't make the Olympic team or win the election to Congress she was just wasting her time, but Donna McAleer doesn't see it that way. In her mind, being a Catholic, Democratic congresswoman from the overwhelmingly Mormon State of Utah made perfect sense and was exactly what the world needed. That was the "reality" she chose to create and believe in. The alternative reality, the "logical" reality that the incumbent male Republican candidate was a lock for reelection, was just a justification for negativity. In her notable 2010 book, *Porcelain on Steel: Women of West Point's Long Gray Line*, Donna McAleer profiled fourteen women who graduated from West Point and went on to remarkable careers (generals in the army, Olympic Team members, business leaders, etc.). Each one of them refused to buy into the status quo and created for themselves a life of their own choosing. Their stories (and Donna McAleer's own story) remind me of what Tiger Woods told Oprah Winfrey after winning his first Masters golf championship at the age of twenty-one (the youngest ever to do so): "The thing I've always believed is that you should never place limits on yourself. Once you do that then that's what you're confined to. Be creative. Go beyond expectations, not just of yourself, but expectations of a human being. Go beyond that, be creative. That's one of the things I've done, that's one of the reasons I've excelled, because I know no boundaries inside."

Great accomplishments like winning the Masters at age twenty-one, writing a groundbreaking book, and sustaining oneself through a grueling congressional election campaign always begin with someone refusing to accept the "boundaries" and the limiting logic of "Do I have at least a 50-50 chance of success?" Use realism and logic when making the small, hour-by-hour decisions, but use some creative fantasy, however "unrealistic" it might be, when thinking about your long-term horizons.

First Victory Alternative Belief #4: The Key Is Enough Knowledge Consistently Applied

"Do more and do it all better" seems to be the operating philosophy when it comes both to personal and team improvement. It's certainly not a bad idea, but it does have its downside, as discussed above. A particular experience I had as a graduate student impressed upon me an alternative—"learn enough and be skilled enough in a couple key areas and then apply those skills consistently." That experience was working for my University of Virginia professors on a study of the determinants of success on the PGA Tour. My job was to manually enter data from the PGA—every stat on every player for a full season—into the university's main frame computer (this was the late '80s, well before today's level of technology). We then asked the computer to analyze the data and determine which stat (average length of drive, number of strokes to reach the green, etc.) was the best predictor of earnings: Which golf stat has the strongest influence on the prize money a player earns? The computer concluded that out of all the available data, the factor that separated the highest earners from lowest earners was only the number of putts a player consistently made per round. The high earners made between one and two fewer putts per round *every round*. They simply did one thing routinely better than the rest. Every player on the tour drove the ball well and every player got onto the green, but the few who putted well *consistently* came home with the big bucks.

I think there's a valuable lesson there. What one or two key points would make the difference in your work if they were consistently addressed? Have these become the focus of your attention every day to the point where you can trust them unconditionally, or are you constantly seeking out new knowledge to expand your tool kit? In his book *Good to Great*, Jim Collins refers to these trusted core competencies as your "Hedgehog," a reference to the old fable in which the sleek and agile Fox tries one trick after another to trap the slow and ponderous Hedgehog, only to be thwarted each time by the Hedgehog's one and only defense—curling up with his bristles extending outward. The Hedgehog can always trust his bristles, just as the best PGA golfers can trust their putting skill. What can you trust?

Special Forces Captain Tom Hendrix decided to trust his understanding of battlefield dynamics during his final deployment to Iraq in 2014, where he advised and trained Iraq's own special forces soldiers. He had been directing a raid by his Iraqi soldiers on an ISIS-held oil refinery for three hours when the radio connecting him to the soldiers on the ground went dead. "Their own commander was dead, the entire troop was pinned down taking heavy fire from an overwhelming enemy force, and my direct line of communication goes out," Hendrix recalled. He stayed in the fight, however, and continued to call in the air strikes that saved the remaining Iraqi soldiers and secured the refinery by watching the video feed from a drone aircraft flying overhead and instantly translating the drone's aerial images into the precise locations where "steel on target" was needed. "I couldn't hear them, and they couldn't hear me, but I could envision the battle space, and I knew the plan, so I had to be confident in it." Later, Tom Hendrix was gratefully embraced by the father of one of the surviving Iraqi soldiers. "My only living son survived, thanks to you!" exclaimed the man, who had already lost two sons in the war against ISIS. "But I didn't do anything heroic," Hendrix observed. "I just used my skills and was confident in them. If you don't believe in yourself, why should anyone else? At some point you decide that you have enough."

Find your core, find your Hedgehog, and put it to work consistently. Maybe you won't be saving lives under fire in Iraq or winning large purses on the PGA Tour, but whatever you do, you'll be in a position to succeed at it.

First Victory Alternative Belief #5: Beliefs Produce Behavior, So Confidence Comes First

The conversation about which comes first, competence or confidence, has dragged on for years, decades, and even centuries (Sun-Tzu's concept of a First Victory dates back to the fifth century BC, so people were talking about it even then). I'm taking the position that the initial spark of confidence (that *First* Victory) has to come first. I have two reasons for taking this position.

First, without that initial spark of confidence, there won't be sufficient energy, drive, motivation, and awareness to develop any competence. We're not necessarily talking about Deion Sanders–level confidence here, just that inkling of possibility, just enough certainty, to try something. As mentioned earlier, you might remember that time in your life when you did not have the competence to ride a two-wheeled bicycle, but somehow you had the idea that you could do it just the same. Something inside you told you that you could develop this very desirable competence, and despite the frustrating falls and scraped knees (evidence that you weren't competent), you maintained enough certainty in your eventual success to motivate you to try it again until you got it. The confidence was there, you just had to feed it with a succession of improvements, a little more distance successfully ridden each time, and that effective thinking enabled your eventual success.

Second, I have witnessed perfectly competent individuals, equipped with all the requisite skills, paralyze themselves into mediocrity and inaction because they refused to allow themselves to be confident. These individuals are very good at telling themselves an interesting story—that whatever competence they might have is somehow less than what is required. They convince themselves that their résumé of success as a high school champion isn't enough to justify some confidence at the college level, or that the professional résumé that earned them their present position isn't enough for them to feel confident that they can get that next promotion. They are nearly always wrong. The truth is that you never know just how competent you are until you act with full confidence. Win the First Victory and all the other ones have a chance to follow.

First Victory Alternative #6: Know Thyself and Trust Thyself, for Every Competitor Is Human and Beatable!

Time to resist the hype, publicity, rumor, and gossip about opponents and competitors. Our media-saturated world provides far too much unhelpful information, and without the proper resistance you might become predisposed to uncritically accept some of that written and broadcast hype as gospel. It's one thing to respect, study, and even learn from an opponent or competitor. That makes great sense, but it's vital to maintain a certain First Victory perspective as you do so. Coaches and managers sometimes unknowingly miss this when presenting the scouting report on this week's opponent to their team or identifying competitors in the marketplace. They highlight the stats, strengths, and accomplishments of that opponent without devoting equal time to describing the weaknesses and vulnerabilities. Here's my recommendation, an alternative way to look at any opponent, competitor, or rival you might have to face, no matter how impressive their hype might be. Use your imagination a little and think about what that person might look like as they shuffle into their bathroom at 6 a.m., eyes still half closed, hair all tousled, yawning and grumbling. What do they look like as they fumble to find their toothbrush or clumsily wash their face? Is anyone impressive when they do that? If you think about what any person looks like as he or she completes the full list of typical morning bathroom behaviors, you'll see them in their most human, most vulnerable, and most beatable condition. This simple reality check illustrates a key point: even the most hyped-up star is a person just like you, and just like you, they come with their complete share of fears, doubts, and imperfections.

It took Kelly Calway, an elite distance runner who has qualified for the Olympic Marathon Trials three times, a while to learn this. She wasn't the top recruit for the distance running group of the North Carolina State University track team, and she let herself get a little too impressed with the blue-chippers. It wasn't until she won the Duke Invitational mile and beat them in her senior year that she realized just how much she had held herself back. As she was training for the 2012 Olympic Marathon Trials and working on her attitude with me she learned to view the top-ranked female runners as no different from her. They had hopes and dreams, just like her. They loved to race and win, just like her. They had fears, doubts, and episodes of injury just like her. This work, in Calway's own words, was "life changing." Instead of lining up at the start of a race and thinking, *Who am I to be right here next to a returning national champion?* she learned to come to the line narrowly focused on her race plan and maintain a steady stream of motivational self-talk throughout her races. Now a US Army major working in the Military Intelligence branch, Calway also coaches the All-Army Marathon team while raising two daughters. And when she races, she respects herself as much as she does any competitor.

So forget the "experts" who create the rankings and the hype. Buying into those images and those stories does nothing for your confidence. Every opponent is human and beatable. Every situation is understandable and winnable.

First Victory Alternative Belief #7: Above All Else, Play to Win

On February 5, 2012, New York Giant quarterback Eli Manning took the snap from the center on his own eleven-yard line with three minutes and forty-six seconds remaining in Super Bowl XLVI. The favored New England Patriots were leading the Giants 17–15, and if the Giants didn't score on this drive, that would likely be the final outcome. Manning first looked to the right side of the field and instantly recognized that his receivers were tightly covered. He looked back to his left, then stepped up and threw a high, arcing pass that covered forty-three yards in the air and dropped perfectly in the hands of receiver Mario Manningham. Two Patriot defenders were right on top of Manningham, but the ball was so accurately placed that either Manningham would make the catch or it would fall incomplete out of bounds. That completion, widely regarded as "the play of the game," set up the Giants' winning touchdown and earned Eli Manning his second Super Bowl MVP trophy.

Two days later, on February 7, Manning took the following question from ESPN radio host Michael Kay on Kay's nationally syndicated radio broadcast: "Do you ever consider the ramifications of failure at a time like that?" referring to Manning's decision to make that throw at that moment. Manning's response was simple and to the point: "That's exactly what you *don't* do." After a second's pause, Manning continued: "You recall all the times you've succeeded in those moments. You recall the game earlier versus the Patriots where you had a fourth-quarter drive to win, you recall the game versus Dallas where you had a fourth-quarter drive to win, Miami, Buffalo. You recall all the successes. And you forget about the games where you didn't have the drives, where you fell short on those opportunities. You have to misremember those moments and just recall the positive ones. That's the feeling."

That's the statement of a performer who has won the First Victory. Rather than operate from the belief that "the team (or player) that makes the fewest mistakes wins, so I better not make a mistake here," Manning operated from the belief that "the team (or player) who plays well despite making mistakes wins." The former belief leads you to worry about both the mistakes you've made and those you might make. That worry leads to tension in the muscles, which in turn degrades your performance, the classic sewer cycle. The latter belief leads to thoughts about playing well *now*, executing the way you've trained and practiced, bringing what you have to the moment at hand. Those thoughts produce Eli Manning's "feeling," which in turn excites the body while not overtensing it so that a great play has the best chance of emerging.

The belief Above All Else, Play to Win and the other alternative beliefs discussed previously are indeed different from what your socialization may have taught you. But let's remember that the purpose of socialization is to encourage conformity and the continued existence of the prevailing social order—its purpose is not to help you discover the full extent of your talents and abilities. If you're a little hung up on thinking differently from the crowd, the very thing that socialization exists to support, you're likely to remain in the crowd. You'll be "normal," but is that really what you want? When my clients admit that they feel a little uncomfortable taking on these new beliefs, that it seems a little weird and strange to be searching for their own version of Eli Manning–level confidence, I always remind them that the difference between special and wonderful and weird and strange is entirely in their own minds, that it's a choice they are making about themselves each and every minute. What choice are you making now? To go with the crowd or to think in ways that help you be ever more special and wonderful? Your First Victory depends on this choice.

CHAPTER SEVEN

Entering the Arena with Confidence

Opening the Vault and Getting Your Money Ready

Tomorrow First Lieutenant Josh Holden will get his shot at a long-sought-after dream; he will try out for a professional baseball team. He will be performing in a competition against top college players for one of a precious few roster spots.

In an hour neurosurgeon Mark McLaughlin will step into an operating room and perform a microvascular decompression of a patient's trigeminal nerve; a surgical procedure that requires him to open a hole a little larger than a quarter in the patient's skull behind the ear, lift up a tiny blood vessel that is pressing on the patient's trigeminal nerve and causing the patient excruciating facial pain, and then place a piece of Teflon the size of a grain of rice between that blood vessel and the underlying nerve.

In an hour First Lieutenant Rob Swartwood will lead his army recon platoon out onto the streets of Fallujah, Iraq, to gather intelligence on the location and strength of insurgent groups. His three six-man squads and six three-man squads will contend with mortar fire, improvised explosive devices (IEDs) placed in their paths, and a local populace that has promised to "make your life hell." And he has to do it under the cover of darkness—daylight just makes his men easier targets.

In fifteen minutes equestrian coach and dressage competitor Christine Adler must step up before her extended family and her parents' closest friends to perform the most difficult task she's ever attempted—delivering the eulogy at the funeral of her recently deceased father.

Each of these former trainees of mine is about to enter a "performance arena." For McLaughlin the surgeon, it's an arena he enters every Monday and Thursday, knowing that what he's about to do will have significant consequences for his patient. For infantry platoon leader Swartwood, and every deployed military leader and first responder, it's an arena he must enter every day knowing that he is quite literally responsible for the lives of other people. For baseball player Holden, and equestrian Adler, it's an arena they are entering once in a lifetime, though for very different reasons. Both of them have entered the competitive arenas of the baseball diamond and the dressage ring hundreds of times before, but this time the stakes are a little higher and nerves a little more alive. And for millions of everyday performers in the working world it's an arena they enter every day to perform the tasks they've chosen as their professional work.

As different as these arenas are, and as different as the motivations and circumstances that have brought these individuals to those arenas are, these "performers" are all opening up their personal mental bank accounts as they enter. Each goes through a transition from their normal state of mind into the personal reality of confidence described in the introduction—a level of certainty about their ability that allows them to bypass or minimize any inhibiting conscious thoughts and execute more or less unconsciously. Winning this First Victory is what allows baseball player Holden to accurately sense and react to each pitch, what allows surgeon McLaughlin to remove tension from his fingers as he delicately maneuvers around tiny clusters of nerve cells, and what allows platoon leader Swartwood to stay cool under fire.

That transition into a confident state doesn't just happen automatically. It happens for a reason, through a deliberate process, a "pregame routine" in the jargon of sport psychology. Having a routine helps you cut through mental clutter and potential distractions and arrive at your arena focused on the present moment and ready to engage in the actions that can lead to success. There are as many different pregame routines as there are performers who use them, but every effective pregame routine includes three key steps: (1) conducting a personal inventory or assessment of yourself, (2) analyzing the upcoming performance situation—what needs to be done, who or what are the competing factors present, and where it will take place, and (3) deciding that you have enough skill, knowledge, experience, and so on to succeed at that moment in that setting. This chapter covers these steps so you will be able to open up your mental bank account during the final moments before any performance, achieve that state of certainty about yourself, and perform with focus, enthusiasm, purpose, and maybe even joy (except when going into combat!). Once you finish this chapter you'll have the tools to become "mentally ready," to transition from *preparing* for the game, meeting, or operation to *delivering* in it; to transition from the process of acquiring skill, knowledge, and competence to the process of releasing that skill, knowledge, and competence. Successfully making this transition is the First Victory that each preceding chapter has been building toward.

An important note to start with . . .

It's not unusual to hear a coach talk about how important it is for their athletes to become "mentally ready." It's also not unusual to hear an athlete on the eve of the Olympic Games or a major golf or tennis tournament mention how their "physical preparation" has been completed and that their "mental preparation" is now their number one priority. Whenever I hear this kind of talk I always flash back to a wrestling clinic I attended years ago when I was coaching middle and high school athletes. The guest clinician was Bobby Weaver, winner of the 48 kg freestyle wrestling gold medal at the 1984 Olympics. After he had led the group of high school wrestlers through a series of technical and conditioning drills, Weaver sat down with them and took questions. "How do you get mentally ready for a match?" asked a youngster, who no doubt expected that the Olympic champion would describe some combination of ritual behaviors that got him all fired up to go out the mat. Weaver's answer, however, was much simpler and much more to the point. He said that the secret to becoming mentally ready to compete is to practice regularly and honestly. There's nothing special to do right before the match to get all "psyched up," because all the real preparation has already taken place through honest daily practice. I would say that same point applies to the First Victory. You win that victory through your daily practice of effective thinking habits over the long haul, and not because you complete some magic ritual right before you step into your arena. As journalist Dan McGinn wrote in the final paragraphs of his book *Psyched Up*, in which he investigates routines, rituals, and superstitions, "There's no substitute for focused practice and lots of it. *Getting psyched up is something you layer on top of actual rehearsal, with the goal of giving you a small boost and an incremental advantage* [italics added]." He concludes, "In our performance-oriented culture, those small boosts can make a big difference." To ensure that you get that final "boost" and put your effective thinking habits to their best use as you prepare to enter your arena, here are three steps you can follow to establish your own pregame routine.

Step One: Take Stock of Yourself—What's in Your Wallet?

Sun Tzu, the author of the classic *Art of War*, put it this way: "If you know the enemy and know yourself, you need not fear the result of a hundred battles." The phrase "know yourself" in this context means conducting a mental inventory of your present capabilities and of your progress, both long term and short term, basically a check on your mental bank account's present balance. Confidence, as we have emphasized from the start, is the total of all your thoughts about yourself, your situation, and all that has taken place in your situation. What is that total right now as you prepare to enter your arena? This is where the journal entries or notes you've kept of your daily episodes of Effort, Success, and Progress come in very handy, as they provide concrete reminders of how hard you've worked, what you've gotten right in your training, and how far you have come. The same is true for any affirmations you may have written; now is the time to read them over again and retell yourself the story that those statements represent.

For a superb example of constructive self-assessment prior to an important performance, consider Billy Mills, the American distance runner who shocked the world in the 1964 Tokyo Olympics, becoming the first American to win gold in the 10,000 m race. Mills had been a three-time NCAA all-American in cross-country, but was virtually unknown in the international running circles when he arrived in Tokyo for the 1964 Games. As he has explained in many interviews and speeches since his gold medal performance, reviewing the entries in his journals and workout books reassured him that he was capable of doing something no one else expected him to do. "I come down to September fifth, six weeks before the games, and once again I put down—'I'm in great shape . . . I'm starting to have a strong finish . . . I'm ready for a 28:25 10,000 meter run in Tokyo . . .'" Mills built up his mental bank account for this one race for over a year, and two days before the Olympics he went back through a whole year of workouts, taking stock of all he had done and all that he had affirmed over that period. His conclusion: "I was totally confident that I could win."

In her keynote address to the 2020 Association for Applied Sport Psychology Conference, another American distance running champion, Kara Goucher, made the very same point about the final preparation for her Olympic Trials and Olympic Games races—going back and reviewing the "confidence journals" she had kept throughout her training to see all the constructive memories and mantras she had collected.

Hall of Fame tennis champion Andre Agassi provides another example in his remarkable autobiography, *Open*. Where Mills and Goucher opened up their bank accounts for review in the form of journals, Agassi opened up his by pulling out memories of previous match wins during his twenty-minute-long prematch shower. "This is where I begin to say things to myself, crazy things, over and over, until I believe them. For instance, that a quasi-cripple can compete in the U.S. Open. [Agassi played in the 2006 U.S. Open Championship despite severe sciatic nerve pain.] That a thirty-six-year-old man can beat an opponent just entering his prime. . . . With the water roaring in my ears—a sound not unlike twenty thousand fans—I recall particular wins. Not wins the fans would remember, but wins that still wake me at night. Squillari in Paris. Blake in New York. Pete in Australia."

Josh Holden's self-assessment leading up to his baseball tryout involves reliving the doubles, triples, and home runs he hit en route to winning the Patriot League batting championship as a cadet at West Point. Mark McLaughlin recalls the hundreds of times he's finished the microvascular decompression procedure and visited a comfortably recovering patient the next day. Rob Swartwood recalls the patrols he led during his first deployment to Afghanistan in the early days of the US military presence there, patrols from which each and every one of his soldiers returned alive and unharmed. These memories are what they find when they open up their mental bank accounts for inspection. Look deeply into your bank account as your performance approaches and really see what is there. If you've been at all diligent in following the guidelines from the preceding chapters, your answer to the question "What's in your wallet?" will be "Plenty!"

Step Two: Take Stock of the Situation—What, Who, and Where

As we have stressed from the start, the degree of certainty you carry into any performance arena will determine how completely you express your abilities, and that degree of certainty is the result of how you think about yourself (hence the importance of the previous step), and how you think about that performance. While Sun Tzu advised "Know the Enemy" in addition to "Know Yourself," we can substitute the word *situation* for *enemy* or the phrase "know what you're getting into" and still get the Chinese military strategist's meaning, acknowledging that not everyone is locked in mortal combat. Once you've inventoried yourself, the second part of a pregame routine is to assess the situation—what is the task to be accomplished, who or what are the opponents/factors that must be considered, and what's the setting or environment that the performance will take place in?

What—The Task

On the surface this seems pretty obvious—there's a game to be won, a sale to be made, an operation to conduct, a test to get a good grade on. But if you give it a little thought you'll discover that to win that game or make that sale you have another task before you—the task of paying attention to what's important minute by minute *during* the performance, rather than thinking about how important the outcome of the performance might be. The soccer player can't see the field and react instinctively if she's constantly looking over her shoulder to see if her play is pleasing the coach, or regretting a missed opportunity earlier in the game, or worrying about how time might be running out. The salesman can't listen to his customer and employ his full array of sales tactics if he's preoccupied with how this particular negotiation might impact his annual quota and any possible bonus in his compensation. While there is unquestionably a desired outcome to be achieved in any performance—the passing grade, the standing ovation, the winning score on the scoreboard—the real task for any performer is to stay focused on the process of performing moment by moment. That outcome certainly matters, and its importance will occupy a certain space in your mind, but your best chance of achieving that outcome isn't by reminding yourself of how important it is, but by keeping all your senses and thoughts tuned in to what's unfolding before you right now.

For Josh Holden and his baseball tryout the real task is settling his attention onto the pitcher. That's the process through which he'll achieve his desired outcome of making the team. For Mark McLaughlin in the operating room the real task is staying patient with each step of the surgical process, without his knowledge of all the potential complications injecting tension into his hands as he operates the microscopes, cutting tools, and coagulators. That's the process through which he'll achieve his desired outcome, to free this patient from pain or return this patient's ability to move. For Rob Swartwood setting out on yet another dangerous patrol, the real task is staying tuned in to the reports coming to him from all his squads and responding with clarity and calmness. That's the process through which he'll achieve his desired outcome, collecting the night's intelligence and bringing all his soldiers safely back inside the wire. Acknowledging the task within your task—the desired process that leads to your desired outcome—is a step in your pregame routine. Whether this is what Sun Tzu had in mind with the phrase "Know the Enemy" I leave to the Chinese scholars, but it's quite possible that the "enemy" he was referring to might have been the human tendency to be so concerned with that desired outcome that the small moments of truth within any "battle" don't receive your full attention.

Who—The Opposing Factors

Scouting your upcoming opponent or researching your sales prospect's needs and history is a standard part of preparation, the most obvious interpretation of "Know the Enemy." But just as knowing the "task" had two meanings, the desired outcome and the necessary process for achieving that outcome, knowing the "opposition" also has an obvious and a not-so-obvious dimension. On the surface athletes compete against one another to win the game and the championship, businesses compete against one another to win customers and market share, and musicians and actors compete against one another for positions in an orchestra, for recording contracts, and for roles on the stage or screen. Beneath this obvious level of "opposition" other opposing forces exist that are invisible, intangible, but no less powerful—the situations in a game, the moments in a trial, negotiation, or operation—that tempt us to abandon our well-deserved sense of certainty and take us into episodes of insecurity. Part of your effective pregame routine should be a brief but honest examination of what could throw you off your game once you enter your arena, and then doing the appropriate Flat Tire drill from Chapter Four to remind you of how you will get back on the road if and when the need arises.

For Christine Adler preparing to deliver her father's eulogy, the below-the-surface "opponent" is that one section of her oration where she tells that one story that always brings her to the verge of choking up. She knows this is likely to happen and she's prepared to pause, settle her breath, and smile when she feels it coming. For Josh Holden preparing to compete against dozens of top college draft picks, the hidden opponent is the pressure he could put on himself by thinking that this tryout is a once-in-a-lifetime, do-or-die situation. He's recognized this potential flat tire and practiced changing it out many times, becoming skilled at telling himself *Here's my chance to do what I love* so he'll remain loose and relaxed. Both these performers understand that their own emotions could compromise their success, so they prepare themselves accordingly before entering the arena. What are the situations in your next performance that might produce the unhelpful emotions of fear, anger, or inadequacy? Bring these opponents out into the light so they can't attack you unawares.

Again, I don't know if this is what Sun Tzu intended by "Know your Enemy," but I do know that the most powerful opponent you may face is not the person in the other color jersey or the one on the other side of the negotiating table. The opponent you might have to prepare most carefully for is the incident that could change you from free and focused to tight and hesitant. Know this enemy and prepare for it!

Where—The Arena Itself

It may only be a myth, but I heard that basketball legend Larry Bird had a pregame habit of dribbling a ball up and down any court he would play on; not just a few times, but long enough and carefully enough to cover every inch of the court so he'd know where every "dead spot" on the floor would be. As a fierce competitor, Bird wanted to "own" every court he played on. He never wanted to be surprised by a soft piece of floor that might throw off his timing as he drove to the basket or pulled up to shoot. Every competitive golfer always plays a practice round on the same course she will compete on, not to refine her swing mechanics, but to become as comfortable as possible with the unique features of that particular course. Hall of Fame hockey player Paul Kariya had a similar practice, seating himself on the bench his team would use hours before game time and envisioning exactly how he would skate and score. When asked why he did that, Kariya replied, "I like to visualize which goal I'll be skating toward." This level of precise mental preparation prompted his college coach Shawn Walsh to comment, "This guy's mind is at a higher level."

I recommend you get yourself "attuned" to each stadium, classroom, operating room, or courtroom where you will perform, well before you actually enter it. Doing so simply gets you a little more comfortable by minimizing the element of surprise and unfamiliarity that is part of every new environment. All my athlete clients are urged to conduct a full "personal orientation" to each new stadium or arena they will be competing in, so that they become as comfortable in that new setting as they are on their home field. The guidelines go like this:

> Walk up into the stands. Go up to the highest seat you can get to and sit down where you have a bird's-eye view of the entire field, court, pool, or rink. Take your time and make friends with the place you'll be

competing in. Identify where in this new arena your bench or seat will be, where you'll enter from, where will you warm up, where the scoreboards and replay screens are. Spend a minute envisioning the opening and closing moments of your performance, and a few other key moments, from that external point of view. What are the spectators going to see you do when they are watching the upcoming game, race, or match? Now imagine that arena packed to the rafters with an excited, energized crowd who is coming to see your opponent perform and hopefully win. How will you handle that noise and that atmosphere? Now walk back down to ground level and go to your bench or team area. Once more look out onto the field, court, or pool and envision your performance, hearing the crowd, and feeling the atmosphere. Make this arena "yours"—a place for you to make a big impact doing what you've trained to do and (hopefully) doing what you love to do.

Had I been advising Billy Mills days before his Olympic 10,000 m race we would both have been sitting high in the stands of that stadium in Tokyo and envisioning each of the twenty-five laps of the race, noting the exact part of the track where he'd surge, coast, and kick to the finish.

For the dancer, the musician, the thespian, the surgeon, the trial lawyer, and the professional speaker, the "personal orientation" drill is the same: get a wide-angle, bird's-eye view of your performance arena—that stage, operating room, or courtroom—and see your desired performance unfold before your eyes in living color and stereo sound.

A version of this personal orientation drill was part of my preparation for my dissertation defense at the University of Virginia, the biggest "performance" of my professional career up to that point. It is traditional for doctoral students at Virginia to conduct their dissertation defenses in the North Oval Room of the historic Rotunda, one of the university's original buildings, designed and constructed by Thomas Jefferson himself and designated as a UNESCO World Heritage site. Students reserve the room weeks in advance, arrive at the designated time, set up their presentation, and hope for the best when the professors arrive. Unfortunately, the students have never stepped foot into that room before, which means that moments before the culminating event of their years in graduate school, they are figuring out where to plug in the slide projector, where to seat their faculty committee, and where they will be standing as they explain their research. Instead of entering the arena with certainty, they are asking themselves and having to answer a multitude of questions at the last minute. Not wanting to have to deal with these last-minute logistical surprises, and since I wanted to deliver my presentation with complete certainty, I arranged to get a sneak preview of the North Oval Room well ahead of my scheduled defense. My "recon" of the room revealed several details of the seating, spacing, and viewing angles that I would have never otherwise been aware of. Like the Canadian diver Sylvie Bernier we met in Chapter Four, who knew where the scoreboard and judges would be as she competed in the final round of the Olympics, I was now able to "see" exactly where each member of my faculty committee would sit, where my presentation slides would be displayed, and where I would be seated as the professors grilled me with their questions. And similar to Bernier's experience of "seeing perfect dives" in that Olympic pool, I "saw" my presentation in the North Oval Room unfold precisely as I wanted it to and prepared myself to take the hard questions that I knew would come. Days later, when it came time for me to deliver before my faculty committee, that room was "mine."

Where will your next performance take place? Will you be comfortably attuned to this "arena"? Perhaps, due to time or travel constraints, you won't be able to conduct an in-person inspection or orientation to achieve that sense of comfort. But it's likely you can find photos of that arena or find a colleague who's been there and who can provide you with some details. Failing all this, your imagination can do the trick—envisioning yourself in that new environment comfortably doing what you know how to do. Any field, stage, court, or conference room can become your favorite if you take the time to let it be so.

Step Three: Decide That You Are Enough—Switch from Saver to Spender, from Workhorse to Racehorse

Now that you have opened up your mental bank account for inspection, and now that you are attuned to your performance arena, the next step in your pregame routine is making a most crucial decision: the decision that you have enough in your bank account. Whether your arena of performance is the ball field, the operating room, the office, or the dangerous streets of Fallujah, you are presented with a choice every time you enter it. Whether you enter that arena every day to perform at a nine-to-five job or enter it once every Sunday afternoon to play professional football that same choice is before you. Will you arrive at the field, stage, or meeting feeling that you *have enough* and that you *are enough* to succeed in this moment? Will you enter your arena having won the First Victory, sufficiently informed about your upcoming task and the demands of the setting, while at the same time sufficiently instinctive to smoothly respond to those demands?

The answer to that question has to be yes. If not, you've made the decision that you are somewhat less than enough, somehow not enough, and doing so results in the doubts that inevitably lead to tension and mediocre execution. The "work" that you have done, be it physical work to become stronger, faster, and smarter or mental work to build up and protect your bank account, only has practical value if you conclude from it all that you're good enough to do what you have to do, against whatever competition or conditions might exist, right there in that particular arena. Making that decision initiates a most crucial internal mental shift—transitioning from an attitude of building yourself up and acquiring skill/knowledge/capability to an attitude of releasing or letting go of everything that you have built up—a transition from being a relatively careful saver to being a relatively carefree (but not care*less*) spender. From being a methodical and dependable workhorse to a spirited and energetic racehorse. At the moment you decide that you are enough, you no longer care about *becoming* better. At this moment what you care about is simply *being your best.*

Michael Phelps, the American swimmer who has collected more Olympic gold (twenty-three gold medals) than anyone ever, and who now in retirement is devoted to expanding mental health services to young athletes, is a case study of this transition. Of all the articles written about Phelps's supremacy in the Olympic pool, the one that stands out to me as a statement on how to enter the performance arena was an August 2008 piece in *Sports Illustrated* entitled "Gold Mind." In it, writer Susan Casey acknowledges that "for all the emphasis on an athlete's body, a large part of Olympic success lies between the ears." She recounts talking with Phelps about competing and noted that as he does so, "his entire energy field changes. *He morphs from laid-back dude into quietly ferocious predator* [italics added]. There is no braggadocio in this, just the same quiet certainty one would expect from hearing Tiger Woods holding forth about chip shots." That "morphing" is what happens when Phelps opens up his mental bank account and prepares himself to release all the speed he's built up through his careful training.

There is no single "right" way to make that transition from saver to spender or from workhorse to racehorse. Christine Adler accomplished it simply by reassuring herself that whatever came from her heart while delivering her eulogy would do justice to her father's memory. Josh Holden made his transition by using the guided envisioning audio tracks we created together during his years playing on the West Point baseball and football teams, narrations that had him imagining "throws that nail base-runners, hits that soar off my bat, steals late in the game that rally the team." Envisioning success in every skill he knew he'd be tested in brought him to a familiar state of excited certainty, to the desired state of informed instinctiveness.

To get his mind "right" for the speed, precision, and intensity needed for success in NCAA Division 1 hockey, 2015 West Point grad Josh Richards used this detailed, prerecorded "Mental Checklist" complete with background music for each section:

"Lose Yourself" (Eminem) Time to bring it in, time to get it going, time to focus in on the game I'm going to play tonight . . .

this is my chance to go all out, to be a force on the ice . . . I start by bringing my mind onto my breathing . . . just feeling the air enter and leave . . . stay with this for a minute, just breathing, letting everything go. . . . . Now see the rink . . . see tonight's uniforms . . . hear the crowd . . . feel the blades cutting on the ice, the stick handling the puck, the excitement, the speed, the intensity . . . Here's how I'm gonna play tonight . . .

"300 Violin Orchestra" (Jorge Quintero) In the D-Zone, I maintain the 5 dice. I take away the wall on the strong side and take away the soft area on the weak side . . . I know where to be in any given situation. . . . I block shots by getting in the lane and getting big. . . . My communication is great. . . . As a team, we create the right transitions and adjustments . . . The ten feet around the goal is our turf and we own it by picking up sticks, blocking shots and clearing the rebounds . . . This is our identity—this is who we *are*!

"We Be Steady Mobbin'" (Lil Wayne) In transition, I get my toes going north . . . On the weak side, I slash . . . and on the strong side, I have poise to make a play . . . I have the vision to pass to our weak side defenseman joining the rush . . . as the puck carrier, I create a scoring chance out of every single odd man rush . . . when I don't have the puck, I drive the middle lane hard and create space for my line mates . . . if there is no offensive play available, I make smart dumps and start the forecheck . . .

"Wild Boy" (Machine Gun Kelly) On our forecheck, F1 comes in hard, cutting the ice in half . . . F2 supports and eliminates the D to D pass . . . F3 has a good read and either hits it or soft locks . . . I want to be the first man in on the forecheck . . . I get great stick on stick to finish the check . . . I reload hard and automatically . . . I create the turnovers that give us scoring chances . . . Defensive guys fear me as I beat them into the corners . . . That's MY puck!

"Going the Distance" (Rocky Soundtrack/Bill Conti) For the power play, my unit is the best at zone entries . . . my job as puck carrier is to be smart and dynamic with the puck . . . I remember "possession over position" . . . I retrieve pucks . . . I have confidence and poise to make a play . . . I keep the puck moving—low to high and high to low . . . I have the best net front presence in the league, getting big and taking away the goalie's eyes . . . when I have the puck around the net, I go to a "happy place" and have the patience of a 50 goal scorer—I bury my chances!

"Lose Yourself." This is how it's gonna be tonight—my best hockey, one shift at a time, right on the edge . . . I set the tone from the opening face-off and make them worry about me . . . I come back to the bench and stay w/my routine so I go out for the next shift in the right mindset—calm yet energized, totally present, absolutely untouchable . . . that's the real me, and that's how I'm playing tonight! Let's go!

To transition to his desired state of certainty as he enters the very consequential arena of a neurosurgery operating room, my longtime client and friend Dr. Mark McLaughlin follows a structured process he calls the Five P's. It starts as he leaves the locker room where he has changed into his surgical scrubs and meticulously washed his hands. Once he's passed through the doorway into the operating room itself he stops and hits *Pause*, taking a moment to completely silence his very busy mind. Like most of us, McLaughlin wears many hats in his life. He's a husband, a father, a youth wrestling coach, and the owner of a medical practice in addition to being a surgeon. Hitting the Pause is how he deliberately and intentionally puts all those other roles into their respective boxes so they won't become distractions during the challenging hours ahead. His Pause can take as little as thirty seconds or can last as long as five minutes, all depending on what's going on in his life on that particular day and how difficult the surgery he's about to perform might be. Standing just inside the operating room with eyes closed, hands at his side, the transition from preparation to performance begins.

Once he's feeling the desired level of calm, his mind goes to thoughts about this *Patient*—"What did they initially come to my practice for? What are they hoping this surgery will do for them? What is debilitating them now? What's gonna be a home run success for them at the conclusion of this procedure? This is the most important moment of their life and they are in my hands."

Next comes the *Plan*. McLaughlin pictures each of the major expected steps in the surgery, starting with the very first incision and proceeding right through to the conclusion: getting sufficient exposure and illumination, placing the proper drainage tubes, inserting the plates, screws, or padding where needed. Thirty seconds of envisioning his Plan adds one more deposit to his mental bank account.

Next he goes into a series of *Positive thoughts*. "You carry with you the wisdom, technique, and experience of Dr. Young and Dr. Jannetta (two legendary neurosurgeons who were McLaughlin's mentors during his residency) . . . It's an honor and a privilege to be here at this moment . . . You were put on earth to do this! . . ." This short reaffirmation brings him feelings of gratitude and power.

Last, McLaughlin offers up a *Prayer*. "Dear God, please help me be the best I can possibly be to relieve this patient's suffering. Give me the strength to navigate through whatever happens in this surgery and give this patient a new chapter in life." This prayer, the final step of his pregame routine, brings McLaughlin to a *state of grace,* where he feels blessed by a transcendent power to bring all his talent, training, and experience to this moment in this operating room with this patient. At that moment, Dr. Mark McLaughlin is *enough*. Once this personal ritual is complete, once that he has achieved that feeling, he opens his eyes, looks at his surgical team, and says, "Okay, let's go." The transition is complete. Mark McLaughlin is finished *preparing* and is now *delivering.* Here is the exact Five P routine that McLaughlin employed while urgently preparing for an emergency surgery on a young woman with life-threatening bleeding inside her brain:

> ***Pause:*** "Mark, close your eyes, go to a quiet place in your mind for one moment and shut everything around you."
> ***Patient:*** "This young woman needs you now, and her parents need you, too."
> ***Plan:*** "Get a drain in first thing. This will buy you some time. Then position her prone fast and get down to the bone quickly!"
> ***Positive thought:*** "You can do this, Mark. This is a moment in your life to make a difference . . . a big difference."
> ***Prayer:*** "Dear God, please help me be my best for Carla. Please help my eyes and my hands to do what is needed to save her. Thank you for this gift you have given me."

In the dangerous world of infantry combat, Rob Swartwood made his transition by first going through all his safety and accountability checklists and then taking time to settle his emotions and achieve, like McLaughlin, an important state of calm. As he explained his process to me, "The thing I did to help myself be ready night after night was surrender my sense of control over the outcome of each patrol. I reminded myself each time that things could happen out there that I couldn't control, and I knew I couldn't let myself get caught up in what the outcome might be, the possibility that me or one of my soldiers might not make it back. I had to let go of concern over the uncontrollables so I could be focused on my role and my actions—the things I *could* control." Once he achieved that focus on his role and actions, Swartwood felt ready to *deliver.*

What About Me . . . ???

What if you're not an Olympic swimmer, a world-class neurosurgeon, or about to head out into a combat zone? What if you're one of the millions of "workday" athletes whose "arena" is the office building, the classroom, or the construction site? Is there a pregame routine for you too? Certainly. The key steps of taking stock of yourself, knowing your situation, and then deciding that you are enough can be followed by anyone who cares to bring out their best when it matters, and for the "workday athlete" it can matter each and every day.

Writer Steven Pressfield provides a valuable example of how to enter into the daily workday grind with a sense of certainty. In his remarkable book *The War of Art* (not to be confused with Sun Tzu's *Art of War*), Pressfield shares how he, and indeed most of us, are confronted daily by a force he calls "Resistance"—the mass of internal distractions, doubts, and fears that keep us from operating at maximum effectiveness. Pressfield knows that force intimately—it's called "writer's block," and he wins the daily battle against it by following his own version of a pregame routine so that he can sit down at his writing desk and produce quality work. It goes like this:

"I get up, take a shower, have breakfast. I read the paper, brush my teeth. If I have phone calls to make, I make them. I've got my coffee now. I put on my lucky work boots and cinch up the lucky laces that my niece Meredith gave me. I head back to my office, crank up the computer. My lucky hooded sweatshirt is draped over the chair, with the lucky charm I got from a gypsy in Saintes-Maries-de-La-Mer for only eight bucks in francs, and my lucky LARGO nametag that came from a dream I once had. I put it on. On my thesaurus is my lucky cannon that my friend Bob Versandi gave me from Morro Castle, Cuba. I point it toward my chair, so it can fire inspiration into me. I say my prayer, which is the Invocation of the Muse from Homer's *Odyssey* . . . I sit down and plunge in."

Pressfield's personal ritual helps him transition from distracted to focused, from tentative to certain. He uses some "lucky" items from his past to aid in the process, but he knows that they are not what makes him confident—they are simply reminders of previous successes and accomplishments—his own way of taking stock in himself. He knows the situation and the stakes—producing the day's pages is the performance he's committed to. And he makes his final transition by opening himself up to a special power that makes him feel that he is *enough*. Now he can sit down and plunge in.

Anyone, in any job, can start their day with a similar personal ritual that brings up feelings of certainty and determination. Find the best in yourself, identify the key tasks to be accomplished, and decide that you have all the knowledge and skill that is required—that you are enough. It's that simple. And that difficult.

Making Your *Decision*

The examples provided earlier show a variety of ways to make that transition from preparation to delivery, from workhorse to racehorse. It doesn't matter "how" you make that decision, it only matters that you make it!

I will acknowledge that this transition is a challenge for many people. The temptation always exists to say to yourself *I sure wish I had studied more/done more/got more practice reps/knew more about this client* as you enter that arena, but doing so only opens the door to self-doubt and puts you on the ineffective sewer cycle of the thought/performance interaction. That cycle is not an option if your intent is to perform at your top level. Similarly, it's easy to ask yourself the question, *Am I ready for this game/exam/meeting/presentation?* but that question also opens the door to a flurry of self-doubt. Better to simply cease and desist with questions altogether once a certain threshold of time or space has been passed. I call this the Statements Only, or SO, Rule—once a certain time has expired, or once a certain line has been crossed, you refrain from asking questions of any kind and make only statements to yourself and to any teammates.

An easy way to understand the SO Rule is think of "game day"—Sundays for the pro football player, Saturdays for college football players, Fridays for most high school football players. Once you wake on that day, once your feet hit the floor and you rise from your bed, you refuse to ask yourself or anyone else any questions, and you deliberately make statements to yourself and anyone else about how well you're going to perform. Even the seemingly innocent questions like "How're you feeling?" or "Ready to go today?" are forbidden, because even those, as simple and innocuous as they might be on the surface, can start up a deeper, more serious, and more negative series of questions taking over your mind. Like a small ember that starts a big fire, those innocent-seeming questions can make you (and your teammates) ask, "Am I *really* ready? . . . Have I honestly done what I should have done to prepare myself for today?" These questions have no functional value when asked on game day. All they do is pull your mental bank account down and dump you into the sewer cycle. Follow the SO Rule and make statements instead, both to yourself and to the people around you who matter—"This is a great opportunity . . . You're gonna get it done today . . . We have chance to do something great right now!"

The SO Rule applies to all of us even if we don't have a designated weekly "game day." Every day that Dr. McLaughlin enters an operating room is "game day." Every nightly patrol that Lieutenant Rob Swartwood leads his platoon into is a singular "game day" experience. Every time writer Steven Pressfield sits at his desk to write the day's pages he is in "game day" mental gear, no longer planning or preparing but "performing." Whenever it's time for you to "perform" in your profession, whether it's several times a day during a nine-to-five day at the office, or nightly as a touring musician, or only weekly as a pro football player, the rule is the same: statements only to yourself and others as you enter your "arena." Let yourself find your own version of Michael Phelps's "quietly ferocious predator" or, if you prefer a less aggressive symbol, a finely groomed and graceful racehorse ready to enter the starting gate and run.

I often get the question "What if I know deep down that I haven't done what I should have done to prepare but I have to go in and take that test anyway?" Answer: go in and take the test with the same 100 percent conviction that you'd have if you'd been the last one to leave the library every night for the last week. Why this answer, when we all know that you can't expect great success without great preparation? Because you never know how much you truly have in your "tank" until you go out and empty it. Didn't study enough? How do you know that for sure? Didn't practice enough? How can you know that until you've gone out and played as confidently and aggressively as possible? Just how much practice/study/preparation is "enough" anyway? You'll never know if you have enough "money" to buy something until you empty all your pockets and put your "money" on the table. Why not open up your bank account and believe that there's enough in it for you to get what you're after?

Chad Allen, managing director for investments at Oppenheimer and Company, didn't start out in the wealth management business armed with an encyclopedic knowledge of market cycles and investment options to impress prospective clients with. What he did have, however, was experience as an army officer and intercollegiate lacrosse player at West Point, and that experience had taught him to focus on what he could control and to make the best use of his personal strengths—in this case honesty, loyalty, and intellect. As he prepared to enter the sales arena and meet a potential client, this twenty-five-year-old rookie knew he couldn't control what his competitors might do or whether the person meeting him for the first time would be put off by his youth. So he aligned his thoughts about the upcoming meeting around what he could control (*I can control the conversation in this meeting today*), and what he could do (*I can get them to agree to a second meeting*). Like Holden and McLaughlin, Allen opened up his mental bank account with a string of positive thoughts as the meetings approached—*I deserve to be in that room because I'm smart and I know how to help families plan their financial lives . . . They deserve to be with ME!* Chad Allen was certainly not the most knowledgeable or experienced financial professional in his firm, but he didn't let that stop him from entering his performance arena with confidence. He certainly hadn't racked up the ten thousand hours of practice and training that are widely considered to be the standard before one can claim to be an expert in anything, but that didn't stop Allen from bringing every bit of expertise he did have to those crucial first meetings.

The trap waiting for everyone is the belief that you can never do too much work or never get too much practice. When you fall for this trap, you enter the exam room furiously reviewing your class notes, or walking into the sales negotiation diligently repeating each line of your marketing pitch in the hope that these last-minute practice reps will magically make up for the deficits in your preparation. The guidance for anyone who might be tempted to wish that they had practiced more, studied more, or learned more prior to stepping into their arena is to acknowledge what you do have, what you have done in terms of preparation, and then close the notebook and declare to yourself, *I'm ready, I'm as prepared as I can be,* I'm enough*! Let's see how well I do with what I've got.*

Helen Maroulis made the statement "I am enough" her mantra as she stepped onto the mat for her Olympic wrestling final against the defending World and Olympic champion, an opponent who had decisively beaten her in their only two previous matches. "That was the most liberating thing I ever said to myself," Maroulis explained in an NBCOlympics.com interview following her victory. "I thought there was some extreme level I would have to reach before I could be an Olympic champion, but you don't have to be extraordinary. You can be an Olympic champion by just being enough."

Are you ready to be *enough*? When you next walk into your performance arena, will you meet the definition of confidence that was set out in the introduction—being sufficiently certain of your abilities to the point where you can perform with the absolute minimum of interference from your conscious, analytical mind? Will you be in a state of "informed instinctiveness" at that moment, having taken stock, first of yourself and then of your situation, and then having decided that what's in your mental bank account is enough for that moment? This is the both the challenge and the opportunity that awaits you.

Conclusion: Let's Imagine That You're Rich . . .

Maybe not Jeff Bezos rich, where you are able to buy an entire country, but rich enough that you don't have to work anymore and that you can afford just about anything you reasonably want. And you got this rich by working hard and saving smart, rather than by inheriting a fortune from a rich uncle or by winning the

Powerball lottery. That means as rich as you are, you're still a little protective of your wealth.

Now imagine that this rich version of you is heading out on a shopping trip for a new car. You've decided on a make and model with certain features and you're on your way to the dealer with a satisfied grin on your face because you know you have enough in your bank account to get the car you want and still have plenty to spare. This feeling of having enough and being comfortable about spending represents a big shift in your attitude. For a long time you worked and saved to build up that bank account, and you were careful with your money so you protected it and invested it wisely. But now your attitude is different; because you are certain about all that you have put into that bank account, you know you have *enough.*

Armed with this different attitude you don't have to worry, as you walk into the dealership, that you won't get what you want or that you'll come away feeling that you didn't get a great deal. Coming in with this considerable emotional strength you get to control the conversation and have the final say.

Isn't that a great feeling?

CHAPTER EIGHT

Playing a Confident Game from Start to Finish

Okay, you've made it to the arena. You went through your pregame routine, took stock of yourself and the situation, and decided that you were indeed "enough." Maybe you're being introduced right now to a packed meeting room or auditorium or you're on the sideline as the national anthem is playing or you're arriving at the fire or accident with your first responder crew. In these situations and a million other situations your confidence and competence are both about to be put on trial, and that trial could be a real bear, because the audience in the meeting room may be skeptical as hell, the opponent on the opposite side of the field (who wants to win just as much as you do) may be tough as hell, and the accident, fire, or battle zone you've just entered may be nothing short of hell on earth. What you're facing right now is not just one "moment of truth" where you need to be in that valuable and effective state of informed instinctiveness, but a whole series of "moments of truth," each of which call for the best you've got. This chapter will show you how to open up your mental bank account over and over throughout the presentation, the game, and the daily grind, so you keep giving yourself another small dose of that *I'm enough* feeling and win one First Victory after another from start to finish.

Let me be honest here. Winning your First Victories from beginning to end of any performance can be a challenge. It would be a little disingenuous for me to tell you that since you have stored up a huge bank account of reasons to believe in yourself and gone through your pregame routine so you enter the arena feeling "rich," your confidence will automatically be at an all-time high as you begin and it'll stay that way as you knock it out of the park minute by minute and play by play right to the end. I'd love to be able to tell you that you'll step up to that microphone and deliver your speech with automatic ease, or rip every drive three-hundred-plus yards down the fairway or listen and respond to the comments and questions from your staff with what seems like animal instinctiveness, easily dismissing any mishap or imperfection. There will indeed be moments where you seem to sail through the speech, the game, the meeting, or the mission with poise and clarity, but there will also very likely be a few rougher moments where you will have to deliberately regain control and reassert your confidence. The very human tendency to focus on the negative (a lasting holdover from our evolutionary past, where this tendency helped our ancestors avoid danger) is alive and well, and it will come knocking on your door no matter how "rich" you are. Add to that the high probability that something or somethings will go less than perfectly as you perform in your arena. The real world of human performance, as we've seen, is an imperfect place, and you are an imperfect human being who is likely to make a mistake here and there. Since you won't be perfect, you'll have to be ready to mentally shrug off those slips and misses as they occur. And then there's the "opponent," either an actual human one, a deadline, or some random but annoying environmental factor like a technical difficulty that throws a monkey wrench at you in midstride or midsentence. Then there's that common socialized belief that during important moments you should "think more" about what you're doing rather than operate from a state of informed instinctiveness. These realities will make it harder to win your First Victories moment by moment, but not impossible. The good news is that no matter how challenging your circumstances of the moment might be, you can *always* choose how you will respond to them, and you can *always* win one First Victory at a time.

During these challenging moments, the methods covered in Chapter Five, "Protecting Your Confidence," will be both your defensive armor and your helpful counterattacking weapons. The mental techniques of treating each imperfect moment of your performance as (1) a *temporary* "just that one time" occurrence, (2) a *limited* "just in that one place" occurrence, and (3) a *nonrepresentative* "that's not me" occurrence will prevent you from sliding into the worrisome "here I go again" trap, the doubtful "I'm messing this whole thing up" vibe, and the debilitating "I stink" state. The nagging voice or voices that intrude on your concentration and further initiate fear, doubt, and worry (the "screaming ninnies" as my colleague Sandi Miller likes to call them) can be silenced one by one by Getting in the Last Word each time they speak up. You calmly *acknowledge* them, firmly *stop* them, and confidently *replace* them with a helpful, task-oriented statement from your mental bank account. And finally, you always have in your personal mental arsenal that paradoxical but very effective Shooter's Mentality, the ability to *look at any miss, mistake, or setback as an indication that you are due to get the next shot, move, or point right*, thinking that your odds for success are only getting better as the game, mission, or workday unfolds. These techniques will protect your sense of certainty from the inevitable attacks by both internal and external opponents and help you to win one little First Victory after another. With these safeguards firmly in place, let's get after a confident performance.

"Mind Without Thinking" . . . "So Alert You Dominate"

Back in the 1960s, psychologist Fritz Perls, the founder of the Gestalt school of psychotherapy, coined the phrase "Lose your mind and come to your senses" to help his patients escape the trap of destructive thinking and lead happier lives. I doubt he was thinking about playing a confident game from start to finish, but his insight was right on target. Any performer in any arena will execute better when they follow Dr. Perls's advice: stop "thinking" so much about what you're doing and pay attention to what is actually happening around you. When I ask any client of mine, whether a competitive sport athlete, a musician, or the manager of a financial services team, about their best, most successful, and fulfilling moments, they always tell me two things that come right back to Perls's recommendation. First, they tell me about feeling "automatic . . . instinctive . . . unconscious," meaning that their decisions and actions, their questions and answers, seemed to happen with very little deliberate thought. In a manner of speaking, they were "out of their minds," meaning they didn't analyze or judge what they were doing while they were doing it; they weren't self-critical, and they weren't, while they were performing, concerned about the outcome or the result. Second, they tell me about being "clear . . . locked in . . . absorbed," meaning their eyes, ears, and feelings, all their "senses," were fully engaged, maybe even turned up a few notches, and that helped them perceive and react effectively to what was going on, whether that was the movement of twenty players on a football field or the slightest movement of the orchestra conductor's baton.

These personal comments on the state of mind experienced during high-performing moments may sound like New Age sentimentality, but they are actually supported by the latest findings in neuroscience. Modern brain scanning and neurobiofeedback technologies have enabled a surge of research into the neuroscience of human performance, and that research supports the idea that a "quiet mind," one that is relatively free of conscious-level analytical thinking is best for high-level execution. Neuroscientists Brad Hatfield and Scott Kerrick concluded their chapter "The Psychology of Superior Sport Performance: A Cognitive and Affective Neuroscience Perspective" in the 2007 *Handbook of Sport Psychology Research* with the statement, "Overwhelming support in the scientific literature exists for the notion that high-level performance is marked by an *economy of brain activity that underlies mental processes*" (italics added). This "economy" refers to the quite literal "quieting of the mind," the shutting down of the brain centers and neural processes that are not required to perform a given task. Hatfield put it very succinctly in a FitTV interview, "From a neuroscience point of view, being totally focused means the brain structures that are essential for performing a task are fully engaged and those structures that are not engaged drop out." With the mind in this quiet, economical state, stimuli from the senses are processed at maximum speed, allowing for faster reactions and smoother coordination of movement. Once my clients finish their descriptions of being "out of their minds" and "into their senses" they nearly always sigh and tell me, "I wish I could be like that all the time."

I wish they could, too, but even if they can't be in that "zone" or experience that "flow state" every time they perform (and I've never met anyone, even among the Olympic champions I know, who is there all the time), they can get closer to it and hit it more often if they take the right steps. You can do the same if you first build up your mental bank account, then decide that you are "rich" enough, and then, before the start of each "engagement" (each baseball pitch, each tennis point, and each check of the production line), deliberately step away from conscious, analytical, make-sure-I-carefully-do-it-the-right-way thinking and open up your eyes, ears, and other senses.

When faced with the greatest personal challenge of his army career, Lieutenant Anthony Randall, West Point class of 1996, chose to get out of his mind and come to his senses. That challenge came at the US Army's Ranger School, a nine-week "suffer-fest," where aspiring Special Operations soldiers and officers carry 120-pound rucksacks while covering up to fifteen kilometers of hilly or swampy terrain. Throughout the grueling days and nights they perform small unit tactical exercises like patrolling, ambushing, and raiding, fueled by only one or two meals a day and getting only three hours of sleep out in the open. The candidates are evaluated daily by a rotating cadre of instructors to see how well they can lead themselves, lead their peers, and ultimately lead a platoon. Whether it's tougher than Navy SEAL training is a question I'll leave for others to debate, but there's no question that Ranger School is a nine-week "performance" requiring a big mental bank account and repeated First Victories. And Randall was in dire need of a First Victory as he neared the end of the final phase of his Ranger School experience. He was far from being in top form when the instructors called him up to plan a nighttime raid and lead a platoon through the raid's execution. He had lost thirty pounds over the previous weeks and his fingers were so cut up he had to wrap them in electrical tape to lift his ruck or hold his weapon. He had already been "recycled" twice, meaning he had failed a phase of the training and had to retake and pass that phase with a new class (only about 30 percent of Ranger School grads make it through without being recycled, and the overall graduation rate is a mere 40 percent). This was Randall's last chance; if this nighttime raid didn't get a thumbs-up from the Ranger School instructors, Randall would fail a Ranger School phase for the third and final time, never get another chance at it, and significantly set back his promising Army career.

Yet despite these difficulties and the reality that he was in a last-chance, absolutely-must-win situation, Randall was completely at peace with himself when the instructors called his number, and he created for himself the needed feeling of control and certainty for the task before him. As he recounted the experience to me, "I was given fifteen minutes to plan the raid out with my platoon sergeant and the squad leaders. I made the decision as soon as I got the order that I would put my mental training to work, applying the skills I had learned at West Point and had been practicing for years"—having that "mind without thinking" (Randall's version of "lose your mind"), and being "so alert that I dominate" (his version of "come to your senses"). With the final words "It's on, dude! We're gonna dominate!" Randall led the platoon out, positioned his weapons squad, selected his assault squads, and split out his last squad for perimeter security. "When the guns went off I was in a state of utter unconscious certainty, completely in control of myself and completely tuned in to every detail as the raid went on. The instructors said it was the best executed mission they'd ever seen." Randall got his Ranger tab, went on to that promising army career, and after years as an airborne jumpmaster and two deployments to Iraq, served as the garrison chaplain at Fort Benning, Georgia.

Making It Routine

To make it easier to get out of your mind and into your senses time and again within a long performance, I recommend using what's known in sport psychology talk as a "preshot routine," *a consistent mental vehicle or path that channels your mind directly onto the task at hand*. That "routine" is something you consistently, habitually do to make sure a desired outcome is achieved. Just as you routinely brush your teeth in order to ensure their appearance and longevity, you can routinely reassert your sense of certainty in your ability to ensure the best possible chance of success throughout your game, mission, or workday. Think of a golfer trying to make her best shot each time throughout her eighteen-hole round, or a quarterback trying to execute at his best on each play of a football game. These athletes have to bring themselves into a state of confidence and focus at least sixty times in every round or game they play, and to help them do that consistently they use a deliberate routine, a short mental process to win a First Victory and get their minds in the right place. Other athletes in other sports with different pauses in the action also have to bring themselves to that same state right before each "engagement"—before each point for the tennis player, before each pitch for the baseball hitter, between each shift for the hockey player, and during each pause for the soccer and basketball player. Even continuous sports like distance running and rowing require periodic reopenings of the mental bank account and the winning of small First Victories to maintain pace and tolerate fatigue. Workday athletes are no different: working a "confident day" from start to finish, slugging through those endless messages in the in-box or monitoring a floor of patients during a twelve-hour nursing shift calls for one little First Victory after another. The preshot routine or pre-engagement routine is like the combination of the padlock or the passcode of the cell phone. Dial it correctly or tap it in correctly and you're "in."

An effective preshot routine accomplishes something very important—it stops you from thinking about the past and the future and puts you right into the present moment, taking you away from overanalysis, needless judgment, and dangerous self-criticism into immediate sensory engagement with what's important *now.* By taking your attention away from all those intruding thoughts and worries, the right preshot routine lets your confidence, your personal sense of certainty, shine through so you can execute without all those thoughts and worries eating up precious mental bandwidth and precious mental energy. In the heat of "battle," whether we're talking about a real combat encounter or the far more common minor battles of the playing field, the concert stage, or the workplace, having a routine that you can go to in order to open up your mental bank account and pull out some confidence is your very best friend.

You can "lose your mind and come to your senses" before each moment of engagement in your performance by following this three step "C-B-A" preshot routine.

1. Cue your Conviction
2. Breathe your Body
3. Attach your Attention

Step One: Cue Your Conviction

If you're a baseball fan you might recall the 1999 movie *For Love of the Game,* in which Kevin Costner stars as an aging major-league baseball pitcher playing a game in a very hostile arena. Throughout the movie, as Costner's character faces one batter after another, he employs a mental routine before each pitch that begins with him telling himself, "Clear the mechanism." That personal, self-initiated statement is his cue to mentally remove himself from the boos and jeers coming from the partisan fans in the stadium so he can focus on delivering one good pitch at a time. That's the kind of short but powerful statement you can begin your preshot or pre-engagement routine with to "cue" your own conviction. The verb *cue* is used here deliberately because it means "to act as a prompt," and in this first step of your routine you are prompting yourself to enter a state of trust and conviction, to bring to the surface your determination to execute well right now.

The use of cue words is a well-established practice in sport psychology, and they have been shown to an effective device for controlling one's attention, controlling one's mood or affect, and as we discussed in Chapter Three, enabling consistent effort. Sport psychology research has demonstrated the effectiveness of cue words for both novice and expert performers in a wide variety of sports (tennis, figure skating, skiing, sprinting, golf, lacrosse, wrestling, basketball, hockey). By using simple cue words or phrases, like "smooth and oily" (for golf), or "explode out" (for sprinting), as they step up to perform, athletes in these studies stayed focused on a helpful action or quality instead of getting tied up in knots thinking about their mechanics or getting worried about the game's outcome.

The research suggests that your preshot or pre-engagement routine should begin with a short, powerful statement. Short, because you are now in the arena and may not have much time; powerful, because the moment calls for the best version of yourself; and a statement because this is Game Day, and on Game Day you follow the Statements Only Rule. Here is a sample of short, powerful cues that my clients have used during their games, races, and performances as their first step:

This is what I do! (NHL player)
Be a Wall! (NCAA lacrosse goalie)
Ease and Fire! (NFL player)
Do it like you know it! (Super Bowl MVP)
Open it up! (NCAA soccer player)

Here's my chance! (Investment advisor)
I've done the work! (Army combat diver candidate)
Time to cruise! (Marathon runner in Olympic Trials)
I'm a First Div Armor guy! (Company commander in a new battalion)

Any statement that brings up or reinforces your resolve to do well will work here. Follow the same guidelines as for constructing affirmation statements: positive wording ("Be a Wall!" instead of "Don't let anyone score"), present tense ("I'm a First Div Armor guy" vs. "I will be a First Div Armor guy"), and powerful ("I've done the work!" vs. "I hope I'm ready"). Be careful also to use a cue that is focused on the process of success rather than the outcome you are striving for; what you need to do or be in the moment instead of the result you are pursuing. "Time to cruise" works better for the marathoner than "Gotta break three hours." Mike Bowman, the longtime coach of swimming superstar Michael Phelps, gave a great explanation of why this difference matters:

"You never want to think 'Man, this is what I've worked so hard for,' because that puts all of your focus on the end result when you need to be totally focused on the process of success. I went to this great class once and they showed these two figure skaters at the Olympic Games. It was one of those years where a Russian girl and an American girl were going to duel to win the gold medal. And they showed the coaches and athletes right before they went out before their last performance which was going to decide who won. The American coach came over to the American skater and said, 'This is what you've worked so hard for,' and the skater was clearly really tight. When you think of it, that focus on the outcome raises your arousal level. If you tell an athlete, 'You've gotta win this event if we're gonna have a chance to win the meet,' that'll raise their arousal level, and sometimes that's appropriate. But this girl was clearly already nervous, and then he says, 'This is what you worked so hard for' and then they tried to high five each other and missed! That's a sign of being too nervous, when you can't perform a fine motor skill. The skater went out and was terrible. Then they showed the Russian girl and her coach, and they were chitchatting about basically nothing at all. He gave a minor technical point, about doing something on the first jump, and she was clearly relaxed, because when you talk about the process it lowers your arousal level . . . that's how we handle our athletes. The ones that are happier, in a better mood, always do better than the ones that are tighter and real serious."

Decide on a cue that brings up your conviction to take a specific action or achieve a particular feeling. You'll know you have come up with a useful cue if each time you repeat it you get that reassuring feeling that you are enough, that you are indeed "rich."

Step Two: Breathe Your Body

Once you've made that powerful statement and felt that reassuring feeling, let it spread throughout your body by taking a deep, comfortable breath or two. Your mother or grandmother wasn't wrong when she told you to "take a deep breath" before starting in on something, but she probably didn't know why it was a good idea or how to actually take an effective breath. Breathing is an activity we generally take for granted, but very few people realize that their breathing can be a powerful performance-enhancing technique. Even fewer make the commitment to learn how to breathe effectively and then incorporate this knowledge into their regular training. Consider just a few benefits of effective breathing as the second step in your pre-engagement routine:

It brings your focus into the present moment—when you breathe purposefully and properly you bring yourself back from both the past and the future and into "right now."

It helps release negativity and self-doubt—a proper breath lets you "blow off" a momentary setback or an episode of simple human imperfection.

It brings up your physical energy level—inhaling fully brings vital oxygen into the bloodstream and a powerful exhale drives out the carbon dioxide that is promoting the buildup of lactic acid.

It reduces the contents of the mind from a crowded jumble of a dozen (a hundred?) competing thoughts down to a sense of unified purpose.

It creates a sense of personal control in difficult situations—when you breathe purposefully and properly, YOU are in control of the situation and not the score, the opponent, or the game conditions.

And if all these weren't enough to make you reconsider the importance of proper breathing as you step up to the line of scrimmage, up to the starting line, or into the negotiation, listen carefully to this statement by karate master Tsutomu Ohshima: "Breathing is the glue connecting the conscious and the unconscious." Ohshima's observation is scientifically accurate; breathing is the only activity that you can consciously control but that also operates without your control, that is, unconsciously. This is because your breathing muscles (more on them in a minute) have a dual set of controls. One set of controls comes from your voluntary nervous system, the one you use when you decide to pick up your cheeseburger or veggieburger. The other comes from your involuntary or autonomic nervous system, the one that operates all by itself to digest that burger and extract and assimilate the nutrients from it once you've swallowed it. This fact has some major implications—it means you can influence a slew of otherwise autonomic and involuntary functions like blood pressure and heart rate through deliberate, conscious breathing. It also means that breathing can influence the transition from conscious mental control of your performance (analytical, mechanical, judgmental, self-critical) to unconscious mental control (automatic, accepting, trusting). So if your intention is to access the power, the skill, the magic that resides in your unconscious, that vast storehouse of capability that you tap into when you "lose your mind and come to your senses," then proper breathing is an important step to take throughout any performance. As Belisa Vranich writes in her wonderful book *Breathing for Warriors*, "Focusing on my breathing means that I can let my body tap into what it knows and has practiced without my brain interrupting."

To perform an effective breath you engage two sets of powerful muscles to move air in and out of the lungs. That's right, muscles! *Breathing is a muscular activity!* Contrary to popular understanding, your lungs do not produce or control breathing, they are merely two large "bags" that either expand or contract in response to the pressure created by your action of your breathing muscles. Expansion of your lungs, which draws air in, is accomplished by the contraction, or tightening, of your *inspiration muscles*—the diaphragm muscle, located between your lungs and stomach and separating your chest cavity from your abdominal cavity, and the intercostal muscles located between your rib bones. When these muscles contract, they enlarge the chest cavity by pressing down on the stomach and intestines (diaphragm) and lifting the ribs outward (intercostals). A correct and effective inhalation is a "down and out" feeling using these muscles, one that makes your belly expand, not an "upward" feeling that lifts the shoulders, the way you were probably taught. Contraction of your lungs, which pushes air out, is accomplished by your *expiration muscles*—the abdominals, both the ones in front and on your sides (the obliques). When these muscles contract (and the inspiration muscles relax), they shrink the chest cavity by pushing the stomach/intestines upward and inward and pushing the ribs inward. A correct and effective exhalation is an "up and in" feeling using these muscles, one that tightens the abs and shrinks the belly. When these two sets of muscles cooperate and work together, the optimal expansion and contraction of the lungs occurs, creating the optimal movement of air in and out.

Try it now, sitting up in your chair or standing comfortably. Begin with an exhale, tightening your abs and your obliques toward your spine and letting air out. Then inhale by relaxing your abs and contracting your diaphragm, feeling your belly expand outward on all sides and your lower ribs lifting as air rushes in. Repeat the cycle comfortably three or four times and enjoy the feeling of energy and ease that it creates. You have now tapped into control over your energy level, your mood, and opened up your unconscious competence by exercising the muscles of inspiration and exhalation. This basic exercise is just that—the most fundamental step in taking advantage of your marvelous human breathing equipment. Every performer, from a weekend golf warrior to an elite tactical military athlete, can benefit from training these breathing muscles as carefully and consistently as you'd train any other muscles. Get to know these muscles and how they work. Your strength, endurance, and concentration will benefit immensely.

When you follow your statement of conviction from Step One with two or three full breaths, finishing with a strong, settling exhalation, you improve your sense of control over the moment at hand. You've decided (going back to the list from Step One) to "Do it like you know it" or "Be the Wall" and your deliberate breathing helps that conviction take hold throughout your body.

Step Three: Attach Your Attention

Time to complete the routine . . . You have essentially "lost your mind" by cueing up a sense of conviction and breathing into the present moment. The usual mind chatter and distractions have been set aside by these two deliberate actions. Now "come to your senses" by attaching your attention to what matters most right now: the ball your tennis opponent is about to serve to you; the final note of the song's opening measure where you come in with your instrument; the spreadsheet in front of you that has the data required for a decision. Attaching your attention to this "target" is the final step of the pre-engagement routine that allows you to perform confidently.

If the phrase "attach your attention" sounds like some complicated psychological process requiring formal study, let me assure you that it is nothing of the sort. It's nothing more than letting yourself become *fascinated* by what you are about to do, letting yourself become so curious, so interested in what is before you, what is around you, and the action you are performing that your senses become utterly absorbed by it all. If you've ever paused for even the slightest second to take in the colors of a beautiful sunset, then you know what I mean by *letting yourself become fascinated* by something (and if you haven't taken in the colors of a beautiful sunset lately, I urge you to do so). This kind of fascination completes the transition "out of your mind" and brings you fully into the here and now of your performance.

Tiger Woods, in his 2004 authorized DVD collection, innocently described becoming so "entrenched," so "engrossed" on the shot he is making that all the background noises and all his self-conscious thoughts disappeared. "It's almost as if I get out of the way . . . I guess my subconscious takes over," he observed, a testament to both "losing" the part of his mind where judgment and self-criticism reside and then opening up his unconscious competence. It's worth noting that at the time of the interviews that made up that DVD collection Tiger Woods was setting a new record for the number of consecutive weeks (264) ranked as the world's number one golfer.

Are you about to receive a tennis serve? Let yourself become *fascinated* by the ball as your opponent tosses it up. Are you about to launch into the next section of your lecture? Let yourself become *fascinated* by your articulation of each word as it comes out. Are you beginning to space out and slow down during your training run? Let yourself become *fascinated* by the steady rhythm of your stride, the sway of your shoulders, or the movement of those all-important breathing muscles.

For a useful variation in becoming fascinated on a target, the story of US Olympic bobsledder Doug Sharp is hard to beat. Some people make an immediate first impression, and Doug Sharp is one such person. At five foot ten inches and 205 pounds, with a blond brush cut, fierce green eyes, and muscles on top of muscles on top of muscles, Doug Sharp is a living superhero action figure. After explaining to him and the other members of the US Army World Class Athlete Program bobsled group who I was and what I did and how I worked, Doug Sharp fixed his penetrating gaze on me and said, "How much time ya got for us?" I soon found out that Sharpie (as he was called) had been a nationally ranked college pole-vaulter who lost his composure and underperformed while trying qualify for the Olympic pole-vault trials. He realized that he had lost his opportunity because he let his mind run away from him, and now that he had a chance to work with a sport psychology pro he wasn't about to let it go.

Over the next fourteen months I spent a lot of time with Doug Sharp and his teammates Mike Kohn (currently the head coach of the US Olympic Bobsled Program) and Brian Shimer (who was training for his fourth Olympic team). Despite facing plenty of adversity, they each learned to look for the best in themselves and build up impressive mental bank accounts. In December 2001 the team of Sharp, Kohn, Shimer, and Dan Steele earned their spot as US Olympic Bobsled number 2.

Fast-forward two months to the XIX Winter Olympic Games. After the first two runs on the first day of competition, sled number 2 was in fifth place, a respectable position for a sled that was not expected to medal. Right behind them in sixth place was the powerful German number 1 sled, a medal favorite.

On the morning of the second and final day of competition, Doug Sharp and I sat down for breakfast at a pancake house right outside the Olympic Village in Salt Lake City, Utah. That evening Doug and his team would have two more runs, two more chances to contend for a medal. Even though no American bobsled team had medaled in the previous forty-six years, I could tell Doug was excited about his team's chances. His mental filter was fully processing the facts that his team was ahead of the number 1 German sled and only .01 of a second behind the fourth-place Swiss sled. As the waitress brought our food I said to him "You guys are sitting pretty right now, Sharpie." Doug glanced over his shoulder, as if to make sure no one was listening, then leaned in over his waffles and omelet. His eyes seemed to grow even brighter, and in a quiet but excited voice he said, "Doc, we're right where we want to be. We're feeling great, and if we get the slightest break, the littlest bit of luck, we're gonna medal!" Right there I knew Doug had won his First Victory.

A few hours later, Doug Sharp and the three other members of USA sled number 2 got the break they were looking for. The driver of the German sled number 1 in sixth place announced that due to a leg injury, he and his sled would have to withdraw. Now the field was suddenly more open, and USA sled number 2 jumped right into that opening. That evening, with the whole world watching, Doug's team reeled off two excellent runs. They got solid push starts and driver Brian Shimer drove like a madman. They took full advantage of that tiny window of opportunity opened up by the German sled's departure and surged ahead of the two Swiss sleds that had finished the first day of competition in second and fourth place. Their bronze medal finish ended a forty-six-year medal drought for US Bobsled and silenced all the critics who had written them off as a no-chance team.

Ever since that conversation at the pancake house I have recommended "looking for your slightest break" as part of any mental preparation and mental toughness training. Looking for that slightest break is an element of the "A" in any C-B-A pre-engagement routine. It plays a useful role in two important ways. First, looking for your slightest break means, by definition, that your attention is focused outward, that you are "into your senses" and scanning your world for anything that might be helpful. Second, looking for the slightest break means that you are acting from a certain degree of certainty that there is indeed a "break" out there that you can use to your advantage once you find it. Think about that: you wouldn't be looking for something if you knew that it didn't exist or couldn't possibly be found. The fact that you are looking for it is a small but significant expression of optimism, one that contrasts starkly with the all-too-common feeling of "I never get a break," or "This just isn't our day." Given the fact that the margin separating victory and defeat, success and failure, is often ridiculously small, the tiniest break or the smallest opportunity can make the biggest difference. But you won't even see it, let alone take advantage of it, unless you are actively looking for it and eager to pounce on it the way Doug Sharp was.

What's the equivalent in your world to the injured muscle fibers in the leg of a German bobsled driver? What little change might qualify as "the slightest break"? A nearly imperceptible nod from a potential customer? A numerical trend revealed on a spreadsheet? The fact that the history teacher loaded the final exam with questions from the one chapter you understand the best? Are you on the lookout for that "smallest break," that "littlest bit of luck"? Don't wait around hoping some big break will come your way. Look for the littlest of breaks and be ready to jump on them when you sense them.

The common misconception that control of your attention, and hence your concentration, in the moment is complex and difficult is just that, a misconception. The truth is that the ability to settle or attach your attention onto whatever is most important at any given moment, to indeed become fascinated by it, is a natural human capability, one that you already have and one that you improve every time you use it. Our modern world of constant distraction, fed by the 24/7 bombardment from social media and online news outlets, has done a very good job of creating the impression that we don't have control over what we pay attention to. But controlling your thoughts and hence your attention is the essence of winning each First Victory. You can choose to take back that control one moment at a time, no matter what is going on around you. Where you put your attention, onto thoughts about what might happen as a result of your actions or onto those actions themselves, is a choice you make minute by minute and engagement by engagement. You can attach your attention and focus your senses onto whatever you choose, onto whatever is most important or helpful at the moment. You can always win the next First Victory by following your own version of C-B-A before each rep of a practice drill, before each heat in the swim meet, and before each meeting throughout the workday.

Nearly every human performance activity consists of a series of engagements, moments where you lock in and deliver separated by moments of recovery and preparation. Football is one such activity, a game of stops and starts where five seconds of fury is followed by twenty-five seconds (sometime a couple of minutes) of reflection and preparation, followed by another five seconds of fury. If you're an offensive line player on the field for a series of plays hopefully leading to a touchdown for your team, your job is to win a First Victory before each snap of the ball no matter what happened on the last play, no matter what happened on your team's previous offensive series, and no matter what happened the last time you played this particular opponent. Over the course of an entire game that means sixty reps of a pre-engagement or "presnap" routine, each one delivering you into your desired sense of certainty. Here's a "Ready-Read-React" routine I've taught to dozens of football players over the years. This *consistent mental vehicle* includes specific "cues" for bringing up conviction, specific "breaths" for settling that conviction, and specific "attachments" for directing the senses to the right target before the snap of the ball.

In the Huddle—Get Ready
- Hear the QB call the play
- "See" the opposing player I'm assigned to hit or control
- "See" the end result of the play (yards gained, or 1st down made)
- Clap Hands at "Ready break!"
- Walk to line thinking *I'm gonna make you pay!*

At the Line—Read
- Take my split, hear the center's call, exhale as I get into my stance
- Check position of opposing linemen and linebackers as I inhale

Exhale as I make blocking call, listen to guard, confirm call

Open up my peripheral vision

Get him in my sights (mentally focus on the opposing player I'm assigned to hit)

Hear the count

At the Snap Count—React

Explode from stance on QB's voice

Hockey is another game of starts and stops, multiple active "engagements" separated by periods of downtime. NHL veteran Danny Brière, whom we met in Chapter Six, would log between fifteen and twenty minutes of total playing time in a game, in shifts roughly forty-five seconds long separated by roughly ninety seconds on the bench. During those ninety seconds he'd follow a "between shift" routine ending with his personal C-B-A signature. Brière's routine, the *consistent mental vehicle* he'd use once he returned to the bench from his last shift, went like this:

Step 1: 10 seconds—Receive comments from coaches about the last shift

Step 2: 10 seconds—"Rinse off" the mistakes, "Rub in" the success

Step 3: 30 seconds—Recharge with 3–5 long slow breaths

Step 4: 30 seconds—Focus on the game, follow the puck, know the game situation

Step 5: 10 Seconds—Fire it up!

*C*ue the Conviction: "Let's Battle!"

*B*reathe and get loose

*A*ttach attention to the ice—"eyes outward"

With the conviction to "battle," a body both energized and relaxed by a quality breath, and his senses focused outward onto the ice, Danny Brière begins his next shift in a state of certainty.

As Dr. McLaughlin the neurosurgeon works his way through an operation, he too engages in stops and restarts, each requiring a reassertion of his confidence and energy. Just as a football game is divided into four quarters, a chess match into opening, middle, and end game sequences, neurosurgeries are divided into steps, and at each step Dr. McLaughlin follows his own version of C-B-A. He pauses, assesses where he is in the overall procedure, then deliberately reasserts what he calls "cognitive dominance," his sense of personal certainty about proceeding to the next step of the operation. That "dominance" is accomplished at each step by a positive thought (C), followed by a full breath (B), and then a deliberate focusing onto the anatomy and equipment that is relevant to the present step in the overall operation right now (A). When you are operating for hours at a time, as McLaughlin often does (his longest is eighteen hours), those short pauses between steps followed by regular C-B-As are what keep you fresh, focused, and certain.

Keeping It Routine When You're Under Pressure

C-B-A is a simple concept, but it can work for you even in the heat and pressure of your toughest, most demanding moments. The key to making it work for you is simply knowing that of all the choices you can make in the moments of truth that you face, the choice to act with a confident mind, a relaxed and energized body, and focused senses always gives you the best chance of doing well. The more your performance matters, the more important it is to for you to free yourself from worry, doubt, fear, and anything else that might interfere with the delivery of the best you've got. The misconception is that the more "important" something is, the more careful, cautious, and "thoughtful" we should be while doing it. But as we have seen throughout this chapter, both the objective neuroscience of high performance and the subjective experience of people engaged in their own high-performance moments argue against that overabundance of thought and in favor of a task-focused state of informed instinctiveness, where you know what you're doing because you practiced enough, and you are relatively automatic or unconscious while doing it.

What takes most people out of their desired state of informed instinctiveness is their misunderstanding of "pressure." We've all heard the statements "Pressure makes diamonds" and "What doesn't kill us makes us stronger" and "You'll succeed when you want success as badly as a drowning man wants air." How often has someone told you that you have to "want it" (the win, the promotion, the success) more than the other guy (or gal)? How many times have you heard the message that if something's important enough to you, you'll find a way to accomplish it? The achievement-oriented world we live in feeds us with endless encouragement, both explicitly and implicitly, to put more pressure on ourselves, to "want it" more, and to inject high levels of personal urgency and importance into our daily tasks and especially into key performances.

Let me offer you an alternative view. Pressure indeed "makes diamonds," but once that diamond is made and it's time for that diamond to shine, you handle it very carefully and put it in a setting where its beauty can be seen. You certainly don't squeeze it any further. It's as beautiful as it will ever be, so leave it alone and just let it shine. Strenuous training does indeed make you stronger, but when it's time to display your strength, you don't want to be tired, distracted, or otherwise compromised (read "pressured"). And that drowning man certainly needs air, but once he gets his head out of the water and gets a little air he can stop breathing so hard. If he continues to hyperventilate long after he's made it to shore, he's likely to lower the level of carbon dioxide in his blood down to where the blood vessels supplying his brain constrict and then he could pass out. So let's establish some perspective here—"pressure" can really help you become better at your sport, craft, or profession when applied in the right places and the right times. That's what training and practice is all about, and to take on that pressure and push yourself sufficiently enough and often enough to develop your capabilities you will indeed have to "want it" pretty badly. But when the time for preparation has passed and the time for delivery has arrived, putting pressure on yourself, getting caught up in how important it is to do well, and "wanting it badly" can all work against you. Why? Because doing so (1) sucks up your mental bandwidth, making it harder to notice what is actually going on in your performance arena, (2) opens the door for worry ("what if I don't . . .), and (3) cranks your emotions into overdrive, creating detrimental muscular tension. The natural engagement of your autonomic nervous system will produce all the energy and focus you need (that's why you have it), so keep yourself on the success cycle and win a First Victory by not injecting too much importance into your performance. Thinking, "Here's my chance to do great/be great/win a big game/nail a big account/have a great time" (you pick it) will serve you far better than thinking, "This is the big one/this really matters/it's now or never/gotta be perfect here."

If you've overstated or overinflated the importance of what you are doing, you will find a potentially dangerous rise in your arousal level, a rise that goes beyond what your body will naturally produce. That rise will create tension, and that tension will interfere with your execution. Perhaps the most important First Victory you ever win is the victory over the temptation to place so much importance on what you are doing, to be so concerned about your outcome that you compromise your ability to execute in the moment. Yes, the game matters, and yes, the interview, the meeting, the negotiation, the concert, and the operation all matter. But the secret to succeeding at something that matters is to keep its importance at the right level. That can mean actually *downplaying* a performance's importance, taking an emotional step back from thinking, "This is what I've studied so hard for or sacrificed so much for," and *taking pressure off yourself.* Wrestler Helen Maroulis put it beautifully in her Vogue.com video "Olympic Wrestler Helen Maroulis Fights Like a Girl"—"I actually have to almost surrender the dream before a match in order to be free enough to wrestle to the best of my ability"—and that's why before a match she's "humming the most positive, happy songs that I know." Neurosurgeon McLaughlin brings himself back into a state of effectiveness during tough, long surgeries by remembering how his mentor Peter Jannetta, a giant of neurosurgery, would keep humming while engaged in the most delicate and dangerous steps of an operation. Legend has it that Hall of Fame baseball manager Casey Stengel would call a time-out and walk out to the pitcher's mound to talk a nervous pitcher back from an emotional edge, like having just loaded the bases, by calmly reminding him that "five hundred million people in China don't care what happens in this ball game." Whether this is a true story or a just a legend, the moral is worth paying attention to: Are you getting so wound up over whether you win or lose, succeed or fail, that you make it harder for yourself to actually execute at or near the top of your ability? Stengel's nervous and worried pitchers could always hit the pause button, "clear the mechanism," breathe and settle, then lock their eyes into the catcher's mitt and throw. So can you. Even if you've dedicated hours a day over years and years to your sport, craft, or profession, and even if your performance has significant consequences for you and your family, you can keep its importance in perspective. How? By acknowledging the fact that doing so will always give you your best chance.

But What If It Really Matters, Doc . . . ?

The more your performance matters, the more important it is for you to free yourself from worry, doubt, fear, and anything else that might interfere with the delivery of the best you've got. That's when you *really* have to step back from thinking about the outcome and the consequences of your performance because the more you focus on them, the more your attention will be divided and the more your power in the moment will be diminished. That's why it's important for ambulance drivers, police officers, firefighters, and military personnel as well as professional athletes on the world stage to internalize a mental routine like C-B-A that helps them open up their mental bank account and find some certainty. The misconception is that the more "important" something is, the more careful, cautious, and "thoughtful" we should be while doing it. But as we have seen throughout this chapter, both the objective neuroscience of high performance and the subjective experience of people engaged in their own high-performance moments argue against that overabundance of thought and in favor of a task-focused state of informed (you know what you're doing) instinctiveness (and you are relatively "automatic or unconscious" while doing it). Reserve your times of being careful, cautious, and "thoughtful" for your preparation, so that you can be carefree (but not careless), decisive, and appropriately "thoughtless" when you deliver.

Conclusion

At the beginning of Chapter One we met Ginny Stevens, the midlevel executive taken momentarily off guard by a last-second directive by her boss to deliver a presentation to a room full of corporate vice presidents. Had she been armed at the time with a well-practiced C-B-A routine of her own (as she is now), she would likely have had a better reaction to that unexpected job tasking than the moment of utter panic she did experience. What might that C-B-A look like?

Cue the Conviction:

I know this product well enough. Be calm and be clear.

Breathe the Body:

Exhale tightening the abs, inhale opening the ribs; exhale and drop the shoulders, inhale and bring in a grateful feeling; exhale and . . .

Attach the Attention:

Scan the room, smile, make eye contact.

Nowadays Ginny Stevens uses variations of C-B-A throughout her workday as she moves from task to task, deliberately winning one small First Victory at a time. Anyone can do this, taking advantage of the natural human capacity for self-awareness and self-control: awareness to know whether you are in the right state of mind to allow your best performance to emerge, and control to adjust your thinking, your feeling, and your sensory settings to win one more First Victory.

CHAPTER NINE

Ensuring the Next First Victory: Reflect, Plan, and Commit—or What? So What? and Now What?

In the journey of one thousand miles,
nine hundred and ninety-nine
miles is only halfway.

—Tsutomu Ohshima

It's over.

The final whistle has blown. The game is over and you're leaving the field.

The applause has faded. The concert is over and you're leaving the stage.

The workday is over and you're headed out of the office.

Wait a second . . . it's not over. There's something you have yet to do and it matters that you do it if you intend to win the next First Victory and perform well at your next opportunity. That something is an honest evaluation of your preparation for and execution of the performance you just finished, what is called in military-speak an AAR, or After Action Review. Until you've conducted a personal AAR and learned everything there is to be learned from the game you just played, the concert you just gave, and the workday you just completed, that game, concert, or workday really isn't over. This chapter will walk you through the steps of an effective AAR so you can extract the maximum value from your last performance and approach your next opportunity with maximum confidence. Don't worry, this is not a difficult process, and you can wait until you've changed out of your game uniform and made it home before you do it. You'll be glad you did it, though, because this honest look at yourself is where you'll find the next deposits for your mental bank account.

An effective AAR proceeds through three steps organized around three questions: (1) *What?* as in what actually happened in the performance, (2) *So what?* as in what can you conclude from what happened, and (3) *Now what?* as in now that you've drawn your conclusions from what happened, what are you going to keep doing, start doing, or stop doing to ensure your next great performance. It can be tempting to rush through these steps or overlook them entirely. We are a driven society and often we drive too fast—once a game, test, or negotiation is over we tend to jump right into the next one without reflection on what we just did, what we might learn from it, and what we might want to do differently. I've advised hundreds of West Point cadets to follow a simple process to prepare for, execute, and then evaluate their various academic papers, projects, and midterm and final exams. Guess which part of that overall process tends to be skipped? Yes, the evaluation part, the AAR, where they could carefully examine where on the test or project they earned points, where they lost points, and what that examination could tell them about their study habits. Don't skip this. Your mental bank account and your resulting sense of certainty will be glad you didn't.

Step One: What Happened?

An effective personal (and team) AAR starts with a dispassionate, nonjudgmental assessment of WHAT actually happened, beginning with the general (overall evaluation of your execution and of your confidence) and moving to the specific (the particular moments when you were at your best and at your worst). This may sound simple and straightforward, but it requires a level of honesty that many people shy away from. This is a time to be both your harshest critic and your very best friend. Both are helpful in different but important ways.

Here are the questions I work through with a client as I conduct the What Happened portion of an AAR. Use these as a guide for what you need to ask yourself to become really clear on What Happened, fitting them into your sport, profession, or application:

1. What was the result (or results)? What was the score or grade or outcome of your performance? While I don't embrace the sentiment that "winning is the only thing" (neither did the author of that statement, Vince Lombardi, who was badly misquoted on the importance of winning), I'm fully aware that results matter.
2. How well did you execute? Look back nonjudgmentally on your "mechanics." What would a neutral observer or a camera record?

3. How well did you maintain the right state of mind? Overall, did you perform with confidence and the right combination of calmness and urgency? In general, did you win the First Victory?
4. How well did you follow your C-B-A routine as the performance went on? How many small First Victories did you win? How much of the time were you fully present and operating out of informed instinctiveness?
5. Where did you slip out of that present, confident state? When you did slip did you pull yourself back quickly or allow these moments to drag out?
6. Where in the performance did you feel like you were really in your "zone"?
7. Where were your highlights? If a video camera had captured every second of this performance, which moments could we edit out to create an ESPN style "highlight reel"? Put this performance through your mental filter and extract the valuable "nuggets"; for example, the top five plays, the three best moments, and so on. This is where you build up your mental bank account!
8. What's the one moment you'd like to have back? The play you goofed on, the most glaring mistake you made. Look at it objectively, acknowledge it, and then forgive yourself for simply being human and imperfect. Once you've learned what there is to be learned from this moment it serves no further purpose and you can let it fade away.

There's a balancing act to be performed when doing this portion of your AAR—a balancing act between self-kindness and self-criticism. That balance is never 50/50—it always shifts a little depending on the situation. If you are reviewing a loss, a bad grade, or an otherwise poor performance, shift the balance in favor of self-kindness—that's when you need it most, but that's also when you are least likely to give it to yourself. It doesn't mean you ignore mistakes and imperfections—far from it. You look at them, too, but instead of dwelling on them exclusively (which is indeed the temptation), you devote MORE time to a careful recollection of whatever small successes occurred during the event. Your selectively perceived "gems" have to outweigh the more easily recognized "junk." To put it in numerical terms, 80 percent of your reflection time should be devoted to reviewing those "gems" and 20 percent to the "junk." Don't think for a minute that this 20 percent isn't "enough" time or energy spent on the "junk." It's plenty, so your mental filter has to be really working at this moment to protect your mental bank account. As we've emphasized throughout this book, using your free will to stay focused on the thoughts and memories of what you want more of, rather than what you want to avoid or are afraid of, is the critical process for building and maintaining your confidence. Unfortunately, we're rather conditioned to only be self-critical, especially after a poor performance or a losing effort. And that typically makes the situation worse rather than better.

The converse is true when reviewing a winning effort or an overall successful performance. During moments such as these (which I hope you have plenty of), it's worth remembering the samurai maxim "After victory, tighten your chinstraps." You'll naturally be in a positive emotional state after your win, so shift the balance of your personal reflection a little and move the emphasis on self-criticism from the 20 percent you used during the losing effort to 40 percent. You'll still be enjoying the warm afterglow of your success (and you should—you earned it), but it's important to manage the all-too-human tendency to ease up on yourself a little after a win.

Whether you are reviewing a big win or a painful loss, the What Happened section of the AAR always concludes with you drilling into your memory the good nuggets, the highlight reel, the moments of effort, success, and progress that build your mental bank account. This takes discipline! This is where the "filtering" techniques from Chapter Two can have their greatest benefit. If you don't finish your What Happened examination with at least a partially renewed sense of purpose and power, then your "experience" is no longer serving to make you better. Gain from every experience you have by employing your precious human free will to identify and take pride in your good moments. Build your mental bank account!

Step Two: So What Does All That Tell You?

The Greek philosopher Socrates pointed out that "the unexamined life is not worth living," so now it's time to move to a deeper level of self-examination, where you take all the facts and events you identified in Step One and make greater sense of them. You just came off the field or out of the operating room or home from the office. You took an honest look at your accomplishments and setbacks. Now examine all that data more carefully . . .

1. What does that information tell you about yourself as a performer right now? What strengths and weaknesses did this performance reveal?

2. What do you know right now that you could not have known before playing this last game, giving this last concert, making this last presentation?
3. And my favorite two So What questions . . . *What are the lessons that this last performance is trying to teach you?* Or *What have you learned from this performance?*

Every answer that you get back from yourself when you ask these questions has immense value. They indicate both the areas where your "game" is solid and dependable, as well as the areas where your "game" needs some attention. In response to these questions, a quarterback looking back on his last game might tell me, "I know I can hit the comeback route . . . I know we can come back from two scores down . . . I know I have to get the ball out quicker when I see a particular defense." A golfer looking back on her last tournament might tell me, "I can read almost any green accurately . . . I still hate playing in windy conditions . . . My wedge play is the weakest part of my game." An executive looking back on his performance during the isolation made necessary by the 2020 COVID-19 virus might tell me, "I handled the initial confusion over our office networking issues well . . . I know we can still serve our clients well even when operating remotely . . . I know I have to take extra care of myself now."

These learnings are the raw materials for your growth as a performer. They remind you of what things you need to *continue doing*, get you thinking about what things you need to *start doing*, and might even tell you about what things you need to *stop doing*. With this knowledge you can be honest with yourself right now about what you need to do to improve or prepare for your next game, presentation, or day at the office. Are there practice drills you need to get in before the next game to address a weaknesses you've identified? Are there topics you need to research or study before your next presentation or meeting? Are there confidence-building drills (e.g., Envisioning, Getting in the Last Word, C-B-A practice) you need do more a little more diligently? Armed with these insights, Step Three becomes simple.

Step Three: Now What Are You Going to Do About It?

You know what happened in your last performance and you've learned the lessons from it. Now what are you going to do with what you've learned? Three things, I hope.

First, take each of those new learnings from Step Two and rephrase them as statements, following the first-person, present-tense, and positive-language guidelines from Chapter Three. For the quarterback it would be, *I hit the comeback route every time . . . We overcome any deficit to win . . . I get the ball out on time against any defense.* For the golfer it would be, *I read each green accurately . . . I hang tough playing in windy conditions . . . My wedge play gets better week by week.* For the executive it would be, *I handle office networking issues well . . . We can still serve our clients well in any conditions . . . I take care of myself so I can take care of others.* These are your affirmations of the day, the stories you can now tell yourself about yourself to build up your mental bank account.

Second, get to work on yourself. Do the drills you need to do, read the research you need to read, study the chapters you need to study to be ready for your next performance. Whether you have a full week to prepare for your next game or only an evening at home to prepare for the next workday, use the time that you have to do what's most important (including getting a good night's sleep). With whatever work you realistically can put in, employ the Immediate Progress Review process from Chapter Two—lock in the memory of your best rep of each drill, allow yourself to feel up-to-date and indeed empowered by each article, chapter, or report you study. The work you put in is important, but what you get out of it and how you feel about yourself once you've done it is vital! Keep building your mental bank account.

Third, envision the success you want next time you are in the spotlight. The final part of your personal AAR is committing to your own success—tomorrow, next week, and in that big, hoped-for performance in your personal equivalent of the Super Bowl or Olympics. You went through the What, So What, and Now What questions in order to enable more success and fulfillment in your life. So go back to the private room you set up in Chapter Four and spend a minute or two seeing, hearing, and feeling yourself experiencing that success: winning that next game, landing that next account, succeeding at that difficult conversation or negotiation. Doing that sets you up to bring a greater sense of purpose, a clearer intent, and some pleasant "intensity" to your life tomorrow. And that is a very valuable First Victory.

EPILOGUE

The Bus Driver, the General, and You

The Bus Driver

Toward the end of my PhD program at the University of Virginia I attended a national conference in New Orleans, Louisiana. Like most graduate students, I would attend conferences to take part in scholarly presentations, network with other students, and pass around my résumé in the hopes of getting a job once I finished my degree. Also, like most grad students, I didn't have a lot of money at the time, and that meant staying at the fancy convention center that hosted the conference was not an option. Luckily, I had a friend in New Orleans who was happy to put me up for the two days, so in addition to doing the presentations, networking, and résumé passing, I got to enjoy the New Orleans food and nightlife with a local guide.

But my best memory of that trip, what has stayed with me for over thirty years, was not the conference itself or the good time I had enjoying jambalaya and bourbon with my friend. It was the bus ride to the conference from my friend's neighborhood, or more specifically, the driver of that bus. On the morning of the conference I waited for the right bus that would take me downtown to the convention center, not expecting to experience anything unusual or memorable. But as I stepped onto the bus the driver greeted me as if I was a long-lost comrade. "Good morning!" he boomed out, his face breaking into a huge grin. "How ya doin'? Great to see ya!" I was no stranger to southern hospitality, but even for New Orleans this guy's friendliness seemed a little over the top. I settled into a seat and took out the folder of term papers I needed to grade, but three minutes later at the next bus stop two more passengers came aboard and were greeted in the same cheerful way: "Good morning! How ya doin'? Great to see ya!" Three minutes later at the next stop the same thing happened: "Good morning! How ya doin'? Great to see ya!" Had this guy swallowed down a little too much coffee this particular morning? When a trio of schoolkids boarded the bus moments later they were greeted just as warmly but also sternly asked, "You got all your homework? Don't you get on my bus without all your homework." The kids smiled and nodded. I guess they were used to this treatment.

As entertaining as the driver's greetings were, the memorable moment came once the bus pulled out onto a bigger street and picked up speed. Looking up into the rearview mirror, and smiling the same huge grin he had greeted everyone with, the driver delivered the following announcement for the entire bus to hear: "Good morning, everybody! It's a beautiful day in New Orleans! I hope you're all having a good day. If you're having a bad day, CHANGE YOUR MIND and have a good day!"

That simple pronouncement of CHANGE YOUR MIND hit me right in the face. Here I was, a graduate student working toward a PhD in the psychology of human performance, and a bus driver had just taught me something every bit as valuable as anything I had ever read in a textbook or scholarly journal—If you're having a bad day, CHANGE YOUR MIND and have a good day! More than thirty years later that voice and that message stills rings in my ears.

Think about how often you have used the phrase "Change your mind" and think about what that phrase actually means. How many times have you heard someone say, "I changed my mind"? How many times have you "changed your mind" and decided, for example, to order the salad instead of the sandwich on the lunch menu, or selected a different pair of shoes to wear from the ones you first thought about? Face it, you "change your mind" dozens if not hundreds of times every day. But do you change it in the way the bus driver was suggesting? Do you change from "I have so much to do, and it's all really hard" to "Let's see how much I can get done" or "I'm stoked to get all that wrapped up and out of the way." Do you change from "Damn, we can't seem to score on these guys" to "Let's keep hammering and dare them to stop us." When you "change your mind" the way the bus driver was suggesting, you step off the sewer cycle of the thought-performance interaction and climb right onto the success cycle. First Victory won.

For a while I was a little embarrassed to admit to myself that maybe that bus driver knew more about confidence and the psychology of success than I did and did more to help people than I did. But then I CHANGED MY MIND. I thought about how great it was that riders on that New Orleans bus route get his cheerful greeting every morning, how great it was that those schoolkids get a reminder to be responsible with their homework every day. And when Hurricane Katrina battered New Orleans in 2005, at least fifteen years after I met him, I thought about how likely it was that that same bus driver, whose name I never knew, but whose face and voice I've never forgotten, was probably helping people get to safety as the city flooded. I imagined him handing out bottles of water in the Superdome building after it was converted to an emergency shelter, or getting an accident victim safely into an ambulance, all the while encouraging everyone to change their minds from despair to hope, to win a First Victory in the toughest of times. Moral of the story: you don't need a graduate degree in psychology to win your First Victories, you just need to be willing to CHANGE YOUR MIND.

The General

I've had the privilege of meeting with, talking to, and conducting formal information briefings for a dozen US Army generals during my years working at West Point, and every one of them is an exceptional individual. You don't get to be a general unless you've got drive, smarts, vision, and the ability to communicate it all. But of all the generals I've met, one stands out—General (retired) Robert B. Brown. That "General" before his name is significant; it means he was a four-star general, the highest rank attainable in the US Army. There are one-star, two-star, and three-star generals in the army, but there are only twelve four-star generals in the entire army. That's twelve four-star generals in command of over *one million* active-duty, Reserve, and National Guard soldiers. General Brown was the 212th four-star general in US history, the first being none other than George Washington himself. Think about what it means to hold the same military rank as the country's first president. That puts you in pretty elite company.

I first met General Brown in the spring of 2003, when he invited my colleague Greg Burbelo and me to Fort Lewis, Washington, to conduct performance psychology training for the officers and NCOs (noncommissioned officers) he commanded in the new Stryker Brigade Combat Team of the Twenty-Fifth Infantry Division. This was the first such training ever delivered to an Army tactical unit, but it was hardly Robert Brown's first exposure to sport psychology. Ever since his days as a high school basketball star in Michigan, Robert Brown had been a believer in visualization and goal setting. At West Point, playing for legendary coach Mike Krzyzewski, he learned how powerfully belief, tenacity, and resilience could transform everyday existence, not just performance on the court, but everything a person did. When he returned to West Point as a captain in 1988 to teach in the Department of Military Instruction, Brown was quickly selected to serve as a trainer in West Point's newly established Performance Enhancement Center, the first designated sport psychology training facility in the country. There he helped create a Mental Skills for Leadership curriculum, and when he left West Point to continue his career in progressively more challenging and important leadership positions, he took that curriculum with him and put it to work. "What I learned about mental toughness at West Point I used every day, in every job, and in every one of my deployments," he recalls, referring to the thirty-eight years he spent as an officer.

When I interviewed him for this book I asked General Brown to share with me his "most confident moment," an incident where his sense of certainty was put to the test, a moment where he had to win an important First Victory. I knew Brown had seen more than his share of action during his two deployments to Iraq so I was expecting a "battlefield story," like the ones I had received from Tommy Hendrix, Rob Swartwood, and Stoney Portis. But instead of a "moment," General Brown gave me something bigger, broader, and more important.

It was December 2004 and the city of Mosul, Iraq, was experiencing over three hundred "incidents" each week—car bombs, improvised explosive devices, and suicide bombers were taking civilian lives every day. Colonel Brown (at the time) and his brigade had the mission of ensuring the success of the first democratic elections, no small task given the ferocity of the al-Qaeda-supported insurgency. Colonel Brown accomplished that mission by working closely with Iraqi forces and implementing a comprehensive plan that included deceptive operations and extensive coordination across the region. By leaking the locations of fake polling sites to suspected al-Qaeda operatives within the Iraqi government; 80 percent of the population voted, and not a single polling place experienced an "incident." But prior to this success al-Qaeda was trying everything possible to defeat Iraqi and U.S. forces and played a trick of its own by recruiting a suicide bomber from Saudi Arabia to join the Iraqi army. On December 21, 2004, in a mess hall packed with American and Iraqi soldiers, that bomber detonated his suicide vest. Twenty-two men were killed and over one hundred wounded. Bob Brown was twenty feet away from the explosion when it happened, and had that bomber stood up before detonating instead of staying seated at his table, Brown would likely not be alive today. As it was, six of the twenty-two casualties were American soldiers under his command, making it the "worst day of my life."

But as horrific as that attack was, Colonel Brown and his soldiers had to go out on missions that very night, and they had to continue executing their most complex and demanding missions for months thereafter. To maintain himself throughout those months, Bob Brown made the deliberate, constant effort to win one little First Victory at a time, "Faith takes practice" he told me, reflecting back on those months following the suicide bombing. "You take that first step, envisioning the success of one mission at a time, and that's how we reduced the number of incidents in Mosul from three hundred a week down to two a week."

"Faith" indeed takes "practice." The notion that faith in yourself or your team comes from a sudden flash of illumination, like the blessing from some fairy godmother in a cartoon movie, is a comforting fiction. If you accept that notion, you'll be waiting for some miraculous intervention and wondering why things never go your way. The truth, as the general points out, is that faith or confidence is a long-haul undertaking that requires practice, practice, and more practice. And you have to put in that practice even when the world hands you a big pile of nasty crap, when part of you screams to just give up and go home. Moral of the story: Winning your First Victories is long-term enterprise. It's a habit that you cultivate, nurture, and *practice, even during the "worst days of your life."*

You!

Decision time . . . Are you willing to CHANGE YOUR MIND as the bus driver advised, and are you willing to do it daily, or hourly, under any and all conditions, as the general advised? If you are, then you can be more confident right now than you have ever been, in any part of your life you care about, and you can gain more confidence every day, no matter what. Your willingness and the tools in these pages are all you need.

> Victorious warriors win first and then go to war,
> while defeated warriors go to war
> first and then seek to win.
>
> —Sun Tzu, *The Art of War*

What kind of "warrior" will you be?
The choice is up to you.

APPENDIX I

Performance Imagery Script Sample

Originally Prepared for
Alessandra Ross—World-Class Track

Note to Reader: This script consists of two parts, confidence-building imagery and specific event preparation and event execution imagery. While it was written for and utilized by a single individual to win her First Victory in the US Olympic Track and Field Trials, it can be modified for nearly any performer and situation.

I've done it, I've made it to the final months before the Trials . . . I've trained well all winter, run some incredibly fast times in both the 800 and the 1,000 . . . I've paid my dues along the way and earned the right to be here . . . Now I am ready to pursue my dream—to go to Sydney as a US Olympian . . . to challenge the best runners in the world . . . I will make that final cut in Sacramento . . . this is my year . . .

I know this will require awesome mental toughness on my part . . . my attitude and the way I use my mind will be the key to my success . . . So right now, and from here on in I commit myself to thinking and feeling like the champion, to totally believing in that 1:56 . . .

I realize that it's a little different up here at the top, that everyone I face will be good . . . but that only gets me excited about showing how good *I* am and what *I* can do when I get the chance . . . I know that I can beat anyone in the country . . . remember Atlanta, how at least two runners in the finals weren't tough enough to go out and fight for a win . . . I was tough enough that day . . .

From here on in I crank up the work ethic, the drive, the motivation, and the desire that got me here to number one in the first place . . . I realize that I can't just take my work habits for granted, that I have to come to practice every day with a purpose, and practice with a purpose . . . if I don't keep the heat on, things will only cool off . . .

From here on in whenever I think about racing, I think about racing great . . . I accept that the best runners in the world will make mistakes—but they do not let it bother them . . . the great runners know that success in racing is all about how you handle your errors, that it's not about being perfect . . . it's about running well despite making errors . . . it is only when we overreact to mistakes that we cause problems . . . I have come so far in dealing with a bad situation and am so much better at it than I was a year ago . . . so I'm going to keep a great attitude and hold my head up between every race, between every practice run, and between every interval . . .

From here on in I refuse to step onto any track without totally and completely deciding to go all out . . . I will bring my best, my most focused attitude with me every time I step onto the track: totally confident, totally focused, totally certain . . . I keep a great attitude, a great mindset during the warm-ups and the moments before the race, totally into the fun, the opportunity, the challenge, and the excitement of racing at the world-class level . . .

I *know* I can hold the pace and I *know* can win . . .

From here on in I accept that it is easier to race great by taking pressure off myself instead of putting more pressure on . . . so I eliminate terms such as "I must," "I should," "I ought to" . . . Instead I just think, *I will race my best now,* and let nothing stop me . . . I know that I don't have to have perfect training to have a great race . . . In fact, I'm so good that I can even be sick for a couple weeks and still turn in a great race time.

From here on in I commit to stay away from anyone, including so-called friends, who want to talk about poor performances, bad weather, or how unfair it all can be . . . my attitude is to enjoy racing and the challenge of racing at the world-class level. . . .

From here on in I commit to racing smart . . . that means no matter what the situation, I keep my mind on winning the next split . . . that means I control my thoughts while on the track and totally trust myself to hold whatever pace I have to . . . racing smart means I keep it simple . . .

From here on in I commit to building my confidence daily . . . I know that I determine my confidence by how I think, how I act, and how I present myself on the track . . . That means I deliberately focus my mind on the thoughts and memories that create energy, optimism, and enthusiasm . . . Like that relay leg at Junior Olympics in North Carolina, when I went after them all and caught them all to win a national title . . . like the time I won Big East by chasing down that girl from Villanova, or the time I caught that girl from James Madison . . . I think about those five state championships in high school . . . about how I beat Jamie Douglas, even with a pulled hamstring . . . I think about passing Michelle w/400 meters to go in that 1,000 at Boston . . . I know I have a great training base and great patience . . . and I am fast enough to beat anyone . . . all I have to do is trust myself . . . the more I keep my mind focused on these strengths and qualities, the more powerful I feel and the more ready I become to go out there and tear them up . . .

As a result of all this *I am more excited than ever* about my next opportunity to go 1:56 in the next big race . . . I will do whatever I have to so I can *trust my speed* the way I *trust my strength* and *trust my toughness.*

I arrive at the track at least ninety minutes before my race . . . as always, I let myself be amused by the carnival aspect of what I see before me—athletes in different colored uniforms, stretching and striding, a mixture of voices and noises that is like no other . . . and as always I feel the delicious onset of race day adrenaline . . . I feel it in my stomach, in my heart and in my legs, the signals that my body is going into a whole new biochemical gear . . . I smile . . . all that power flowing through my body right on cue, getting me ready to take it to another level . . . I know that other racers are getting nervous, too, but that they are wishing the butterflies and jitters would go away . . . not me, I know they are signals of power . . . I have waited for them and looked forward to their coming . . . now they are here, and I love it . . . My goal today is to race with total reckless abandon . . . nothing can stop me from throwing all the fear away, totally trusting all the training I've put in and just going for it . . . whenever I do that, I know I am the winner, regardless of where I finish . . .

I get my number, check the heat sheet, and evaluate the

competition . . . the names are mostly familiar, I think about what strategy will beat them and lock it into my mind . . . turning away from the official's station, I eliminate all the elements of doubt one by one . . . I *know* I can win . . .

Now I begin my easy stretching in the warm-up area . . . I take off my shoes and work out my hamstrings, my hips, and my back . . . I smile and talk with friends and training partners as they come and go . . . Other racers come by . . . I am friendly and at ease with them . . . just like me, they love to race and win . . . The few racers I don't care for I can easily ignore . . .

With sixty minutes to go before the race I start my warm-up run, ten or fifteen minutes at a nice slow pace, enough to break a sweat and let everything get loose . . . As I surrender to the comfortable pace I see the race in my mind, and hear the splits being called . . . my legs feel loose and warm, ready to explode, ready to win . . .

With forty-five minutes to go I find a quiet spot where I can enter my own little world and stretch with real focus and real intensity . . . I think about how much fun it is to race, how lucky I am to be here and have this chance to let it all hang out . . . each muscle lengthens and loosens as I breathe deep and let go . . . in my mind I see my strategy unfolding and feel myself flying down the last 200 with perfect form, blowing by them all in the final stretch . . .

It's twenty-five minutes before race time now, and I feel the excitement building as I breathe long and deep . . . I go into my butt kicks, my high knees, and my knee lifts . . . I feel my focus drawing in as I pace to the ends of the warm-up area . . . If my shoulders feel tight I stretch them . . . Quick and strong is what I feel . . .

Now there are fifteen minutes to go and it's time to put on the spikes . . . as I tighten down the laces, my heart rate picks up and my focus becomes tighter . . . I check in with the officials and attach the hip number . . . time to start the strides . . . I feel the smooth form as I stride, relaxed and powerful . . . I love this feeling . . . In my mind I am passing one opponent after another, untouchable in the last 200 . . .

Now there are ten minutes to go and I bring the knees up in rapid fire . . . I have never felt better, stronger, faster, more ready to go . . . only a little while to wait . . . now I pace and relax with five minutes to go, thinking about where I want to be, ready whenever they call me . . .

In the holding pen I let my body stay loose . . . All the work is done . . . this is what I'm here for . . . I can't wait to call Dad and tell him how great I did . . . I keep my sweats on until the last minute, then shed them just before stepping onto the track . . . as my feet touch the track surface I'm thinking, *Yes! This is what I love!* . . . I stride up and back until at last I hear it—**"Final call for the 800. All runners report to the start line"** . . . Yes! Now I get to do it, now I get my fun! . . . I feel only eagerness, excitement, and confidence as I get my lane assignment and walk over to it and see the line I will take when I cut in . . . I think, *Get out fast!* and picture the great start I'm going to have in this lane . . . I feel eager, ready, and totally happy . . . **"Runners set"** . . . my mind goes blank as I step in with my right foot . . . **BANG!**

I go out hard and cut in perfectly, getting right behind the leader and splitting her shoulders . . . now I settle in, adjust my breathing and relax . . . drill that hole in her back and stay with her . . . breathe, stride, and relax . . . breathe, stride, and relax . . .

breathe, stride, and relax . . . My first 200 is 29:30, "Great, right on target" . . . I can go forever at this pace . . . I let her do all the work and just ride her, just ride her . . . breathing big, loose arms, relaxing and having fun all the way, relaxing and having fun . . . my 400 split is 58, still on target . . . now the race really starts . . . I feel the others falling off the pace, but I stay focused on that hole I have drilled into her shoulders, eyes front, head still, just burning it up . . . whenever she moves, I move with her, keeping the gap . . . This is what I love, this is what I came for, the chance to feel my speed and strength come out . . . At 600 meters I come through at 1:27 and I have never felt more alive or more powerful . . . I have her right where I want her . . . with 150 meters to go I take it out and go for the tape . . . smooth fast turnover . . . loose arms and fiery breath . . . I surrender to my speed, letting it take me faster than I have ever run before . . . I am totally consumed by the joy and the speed, my feet hardly touching the track . . . I burst through the tape with my final surge and win the race at 1:56 . . . Yes! I did it!

This is where I choose to be, this is what I love . . . the challenge of going up against the best and dominating . . . feeling that rush that only comes from quality competition, . . . And if I really love it, that means I have to love it all, not just when things are going my way . . . Sometimes it *will* seem too hard, but if it wasn't hard, then anyone could do it . . . Whenever it gets tough, I just remind myself that I love it . . . And when the Trials come I will be totally ready to let it all hang out and surrender to my speed . . . Nothing can stop me when I decide to go! . . . This is my year!

APPENDIX II

After Action Review Worksheets

Step One: What Happened?

What was the result (or results)? What was the score or grade or outcome of your performance?

How well did you execute? Look back nonjudgmentally on your "mechanics." What would a neutral observer or a camera record?

How well did you maintain the right state of mind? Overall, did you perform with confidence and the right combination of calmness and urgency? In general, did you win the First Victory?

How well did you follow your C-B-A routine as the performance went on? How many small First Victories did you win? How much of the time were you fully present and operating out of informed instinctiveness?

Where did you slip out of that present, confident state? When you did slip, did you pull yourself back quickly or allow these moments to drag out?

Where in the performance did you feel like you were really in your "zone"?

Where were your highlights? If a video camera had captured every second of this performance, which moments could we edit out to create an ESPN-style "highlight reel"?

What's the one moment you'd like to have back? The play you goofed on, the most glaring mistake you made. Look at it objectively, acknowledge it, and then forgive yourself for simply being human and imperfect.

Step Two: So What Does All That Tell You?

What does that information tell you about yourself as a performer right now? What strengths and weaknesses did this performance reveal?

What do you know right now that you could not have known before playing this last game, giving this last concert, making this last presentation?

What are the lessons that this last performance is trying to teach you? What have you learned from this performance?

Step Three: Now What Are You Going to Do About It?

Take each of the new learnings from Step Two and rephrase them as statements, following the first-person, present-tense, and positive-language guidelines from Chapter Three.

1.

Get to work! List the three most important actions you need to be ready for your next performance. Be realistic about it—what can you do in the time you have?

As you put in that work and preparation toward your next performance, lock in the best memories of it (E-S-P). Keep building your mental bank account.

Envision the success you want next time you are in the spotlight. Go back to your private room and spend quality time seeing, hearing, and feeling yourself experiencing that success.

Acknowledgments

Just as no person is an island unto him or herself, no book is a result of one person's mind. This book came about through the efforts and influences of many others, and they deserve acknowledgment here.

I begin with the three people who were instrumental in the book's production: Peter Hubbard, my editor at William Morrow, and his great team, who pulled all the necessary levers, (including me), to create the finished volume; Lisa DiMona, agent extraordinaire, who brought my proposal from obscurity into the market's light; and Linda Carbone, my patient freelance editor, who took my disorganized musings and molded them into a marketable proposal.

My most sincere thanks to five individuals who had the greatest impact on my work as a sport psychology professional.

Miller Bugliari, teacher and coach at the Pingry School, gave me my first glimpse into the mind's role in performance, even though I never played on any of his championship soccer teams.

The late Colonel Bob Rheault, West Point class of 1946 and one of the original U.S. Army Green Berets, was my watch officer at the

Hurricane Island Outward Bound School in 1971 and served as a life-changing model in how to live a life guided by challenge and compassion.

Mr. Tsutomu Ohshima, the first karate master to teach in the United States and leader of a worldwide network of traditional karate practitioners. The origins of sport psychology go back to the mental training of warriors all the way to the *Bhagavad-Gita,* and Ohshima-sensei is the living embodiment of that training. No warmer, humbler, yet more powerful individual lives on this planet, and it is my great honor to be one of his black belts.

My sport psychology mentors at the University of Virginia are the real authors of this book. Dr. Bob Rotella took me under his wing for four years as his grad assistant, teaching me that the psychology of success was a personal choice and showing me how to coach others to achieve it. I can never thank him enough.

Dr. Linda Bunker, my other University of Virginia mentor, pulled me into her office one day and asked me if I would edit, update, and rewrite with her a chapter on "Cognitive Techniques for Building Confidence and Enhancing Performance" for the textbook *Applied Sport Psychology: Personal Growth to Peak Performance.* That chapter and its successors launched my literary forays into the topic of confidence.

Then there are a few people who would not let me rest until I had followed through on my intention to write this book. Dr Mark McLaughlin, my longtime client, friend, and tireless supporter, told me one day years ago, "There's a book that only you can write, Nate." And he kept on me to write it. My dear friends Gerry and Sandie Rumold kept up a constant stream of support through the months and years, chapter by chapter, until it all came together. My Outward Bound buddy Marty Aaron kept up a similar steady stream of encouragement all the way from Texas.

Next are the clients, advisees, and students who shared the stories that appear in these pages. Without their willing contributions this book would lack authenticity and meaning. In order of their story's appearance, they are: Eli Manning, Stoney Portis, Jill Bakken, Connor Hanafee, Max Talbot, Ginny Stevens, John Fernandez, Bobby Heald, Anthony Stolarz, Allesandra Ross, Phillip Simpson, Gunnar Miller, Kevin Capra, Paul Tocci, Jerry Ingalls, Dan Browne, Mario Barbato, Maddie Burns, Joe Alberici, Nick Vandam, Jonas Anazagasty, Danny Brière, Donna McAleer, Tommy Hendrix, Kelly Calway, Josh Holden, Mark McLaughlin, Rob Swartwood, Christine Adler, Josh Richards, Chad Allen, Anthony Randall, Doug Sharp, and General (retired) Robert Brown. My sincere thanks to everyone on this list.

I would be equally remiss not to salute and otherwise acknowledge the hundreds of other clients, advisees, and students whose stories could have appeared in this book. I thank them all, too, for letting me share in their athletic, military, and professional pursuits.

To the many coaches across the nation who allowed me to train and counsel their athletes over the years, please accept my thanks. I owe a huge debt to Tom Coughlin, Mike Sullivan, Kevin Gilbride, Pat Shurmur, Ken Hitchcock, Johnny Stevens, Peter Laviolette, Bob Sutton, Jack Emmer, Joe Alberici, Bob Gambardella, Paul Peck, Doug van Evern, the late Tod Giles, Chuck Barbee, Joe Heskett, Jim Poling, Russ Payne, Kevin Ward, Brian Riley, Kristin Skiera, and Web Wright.

It has been the honor of a lifetime to teach sport psychology at the United States Military Academy and serve the Corps of Cadets. I have shared that honor with many wonderful colleagues, both active-duty military and civilian, who have either worked directly in, or worked to support, the Performance Psychology Program. They are: Sandi Miller, Jen Schumacher, Kat Longshore, Jeff Coleman, Angie Fifer, Bernie Holliday, Dave Czesniuk, Greg Bischoping, Darcy Schnack, Doug Chadwick, Seth Nieman, Carl Ohlson, Jim Knowlton, Pete Jensen, Travis Tilman, Greg Burbelo, Pierre Gervais, Bruce Bredlow, George Corbari, Larry Perkins, Jeff Corton, Brad Scott, Rich Plette, Thad Weissman, and Bill McCormick. None of those people, however, would have had said privilege were it not for Louis Csoka, who established a fledgling sport and performance psychology program at West Point and then hired me to make it fly.

And finally, because she is both first and last in my heart, I thank my wife of thirty-nine years, Kim, who in the words of Bob Dylan, has been there for me both "in the darkness of my nights and in the brightness of my days." Let's go walk in fields of gold.

Reference Notes

Preface

xiv **"becoming a stronger, more":** Personal communication with New York Giant head coach Tom Coughlin, February 2007.

xiv **"This is a different":** Statements made by Fox Sports TV analyst Daryl Johnston on December 9, 2007.

Introduction: What Confidence Is and Isn't

2 **"There I was in":** This and all statements and comments by Lieutenant Colonel Stoney Portis from personal interviews conducted July 2020.

6 **"I'm a very modest person":** Drew Brees, interview with Steve Kroft, *60 Minutes,* CBS, September 26, 2010, https://www.youtube.com/watch?v=kHxsWJW_OgI.

8 **"When we focus too hard":** Harper Lecture with Sian Beilock, "Perform Your Best Under Stress," May 3, 2015, Hilton Garden Inn, Minneapolis, MN, uploaded to YouTube June 16, 2015, https://www.youtube.com/watch?v=nuH6X0Tx--I.

13 **"We just had confidence":** Comment by Jill Bakken in CBS TV coverage of the 2002 Winter Olympic Games, February 2002.

16 **Bob Rotella, the sport psychology:** Bob Rotella, *Your 15th Club* (New York, Free Press, 2008), 6.

18 **"The thing that haunts":** Peter King, "Who Let This Dog Out?," *Sports Illustrated,* January 29, 2001.

21 **"It was quite complicated":** Comments by Ilya Kulik in CBS TV coverage of the 1998 Winter Olympic Games, February 14, 1998.

25 **"the last of the human freedoms":** Viktor Frankl, *Man's Search for Meaning* (Boston: Beacon Press, 2014), 66.

25 **"Every day, every hour":** Frankl, *Man's Search for Meaning,* 66.

Chapter One: Accepting What You Cannot Change

29 **Conversely, other studies showed:** Redford Williams, *Anger Kills* (New York: Penguin Random House, 2012).

35 **"The film can be perceived":** Jadranka Skorin-Kapov, *Darren Aronofsky's Films and the Fragility of Hope* (New York: Bloomsbury Academic, 2015), 96.

35 **Kate Fagan's 2017 book:** Kate Fagan, *What Made Maddy Run: The Secret Struggles and Tragic Death of an All-American Teen* (New York: Little Brown and Company, 2017).

36 **Research shows that the highest:** Joachim Stoeber et al., "Perfectionism and Competitive Anxiety in Athletes," *Personality and Individual Differences* 42, no. 6 (April 2007): 959–69; Melissa Dahl, "Alarming New Research on Perfectionism," *The Cut*, September 30, 2014, https://www.thecut.com/2014/09/alarming-new-research-on-perfectionism.html.

37 **"Those athletes who strive":** Stoeber et al., "Perfectionism and Competitive Anxiety in Athletes."

38 **"but that's the irony":** Joe Flower, "Secrets of the Masters: It's Not Just Technique That Makes These Athletes Extraordinary—It's the Attitude and Commitment They Bring to Their Performance," *Esquire*, May 1, 1987.

43 **when you're "nervous":** Simon Sinek, "Nervous vs Excited," YouTube, May 16, 2018, https://www.youtube.com/watch?v=0SUTInEaQ3Q&t=2s.

44 **"Ward then headed to":** Skip Wood, "Easygoing Ward Battles Nerves on March to MVP," *USA Today*, February 7, 2006.

44 **"Definitely, my heart was pounding.":** Michael Johnson, interview with Bob Costas on NBC TV, August 1996.

46 **"You want to go out there":** "Does Bill Belichick Ever Get Nervous? 'Every Week,'" CBS Boston, January 18, 2019, https://boston.cbslocal.com/2019/01/18/bill-belichick-patriots-chiefs-afc-championship-game-nfl-nerves-nervous/.

46 **"Before an Op, everyone feels":** Richard Marcinko and John Weisman, *Green Team: Rogue Warrior* (New York: Pocket Books, 1995), 355.

49 **"At this point, there's":** George Leonard, *Mastery: The Keys to Success and Long-Term Fulfillment* (New York: Penguin Books, 1991), 16.

50 **recent neuroanatomy research:** J. Pujol et al., "Myelination of Language-Related Areas in the Developing Brain," *Neurology* 66 (2006): 339–43; F. Ullen et al., "Extensive Piano Practicing Has Regionally Specific Effects on White Matter Development," *Nature Neuroscience* 8 (2005): 1148–50; T. Klingberg et al., "Microstructure of Temporo-Parietal White Matter as a Basis for Reading Ability," *Neuron* 25 (2008): 493–500; B. J. Casey et al., "Structural and Functional Brain Development and Its Relation to Cognitive Development," *Biological Psychology* 54 (2000): 241–57.

51 **"myelin serves as the insulation":** Dan Coyle, *The Talent Code* (New York: Bantam Dell, 2009), 32.

52 **"To love the plateau":** Leonard, *Mastery*, 49.

Chapter Two: Building Your Bank Account #1: Filtering Your Past for Valuable Deposits

56 **"The last thing I need":** Richard Hoffer, "Fear of Failure; His Lifetime Average Is .335 and Climbing!—But as Tony Gwynn Zeroes in on a Sixth Batting Title He Still Hones His Stroke and Obsesses Over His Videotapes," *Sports Illustrated*, September 13, 1995.

57 **"I decided right there":** Comments by Captain (retired) John Fernandez, personal communication, 2018.

58 **whatever your conscious mind:** J. A. Bargh, ed., *Social Psychology and the Unconscious: The Automaticity of Higher Mental Processes* (Philadelphia: Psychology Press, 2006).

69 **"stopped a two-on-one breakaway":** Personal communication from Anthony Stolarz, 2015–2016.

72 **the science tells us:** Julija Krupic, "Wire Together, Fire Apart," *Science* 357, no. 6355 (2017): 974–75, doi:10.1126/science.aao4159.

75 **"Is your present way":** Bob Rotella, *Your 15th Club* (New York: Free Press, 2008), 8.

Chapter Three: Building Your Bank Account #2: Constructive Thinking in the Present

78 **"a 19-lesson course":** West Point Military Movement Course, https://www.westpoint.edu/military/department-of-physical-education/curriculum/military-movement. (Accessed 15 March 2020.)

81 **"a belief or expectation":** Michael Biggs, "Prophecy, Self-Fulfilling/Self-Defeating," in *Encyclopedia of Philosophy and the Social Sciences* ed. Byron Kaldis (Thousand Oaks, CA: SAGE Publications, Inc., 2013).

81 **"If men define situations":** W. I. Thomas and D. S. Thomas, *The Child in America: Behavior Problems and Programs* (New York: Knopf, 1928), 571–72.

82 **"Our life is what":** Marcus Aurelius, *Meditations* (New York: Dover Publications, 1997).

82 **"A man is what":** Ralph Waldo Emerson, *Selected Essays* (London: Penguin American Library, 1982).

83 "an overarching narrative": G. L. Cohen and D. K. Sherman, "The Psychology of Change: Self-Affirmation and Social Psychological Intervention," *Annual Review of Psychology* 65 (2014): 333–71, doi:10.1146/annurev-psych-010213-115137. PMID: 24405362.

84 "Self-affirmation can thus lead": D. K. Sherman, "Self-Affirmation: Understanding the Effects," *Social and Personality Psychology Compass* 7, no. 1 (2013): 834–45.

84 "It is clear that": Alia Crum and Ellen Langer, "Exercise and the Placebo Effect," *Psychological Science* 18, no. 2 (2007): 165–71.

95 "So we did all": Alexander Wolff, "Whooosh! To the Delight of Their Families and Fans, Dan Jansen Won at Last, and Bonnie Blair Won Again," *Sports Illustrated*, February 28, 1994, https://www.si.com/vault/1994/02/28/130564/whooosh-to-the-delight-of-their-fans-dan-jansen-won-at-last-and-bonnie-blair-won-again.

98 "Who are you NOT": Marianne Williamson, "Our Deepest Fear" in *A Return to Love: Reflections on the Principles of A Course in Miracles* (New York: HarperCollins, 1992).

101 "People develop new understandings": Richard Tedeschi, quoted in Lorna Collier, "Growth After Trauma," *Monitor on Psychology* 47, no. 10 (2016): 48, https://www.apa.org/monitor/2016/11/growth-trauma.

Chapter Four: Building Your Bank Account #3: Envisioning Your Ideal Future

109 "I could feel it": Bruce Newman, "At Long Last," *Sports Illustrated*, September 19, 1991.

112 Recent advances in magnetic: Kai J. Miller et al., "Cortical Activity During Motor Execution, Motor Imagery, and Imagery-Based Online Feedback," *Proceedings of the National Academy of Sciences* 107, no. 9 (2010): 4430–35; J. Grèzes and J. Decety, "Functional Anatomy of Execution, Mental Simulation, Observation, and Verb Generation of Actions: A Meta-Analysis," *Human Brain Mapping* 12, no. 1 (2001): 1–19; H. Burianová, L. Marstaller, P. Sowman, et al., "Multimodal Functional Imaging of Motor Imagery Using a Novel Paradigm," *NeuroImage* 71 (2013): 50–58; M. Jeannerod, "Mental Imagery in the Motor Context," *Neuropsychologia* 33, no. 11 (1995): 1419–32.

112 your imagination stimulates: T. X. Barber, "Psychological Aspects of Hypnosis," *Psychological Bulletin* 58 (1961): 390–419.

112 "Imagery, or the stuff": J. Achterberg, *Imagery in Healing* (Boston: Shambhala Press, 1985), 3.

113 **In one of the:** Edmund Jacobson, "Electrical Measurements of Neuromuscular States During Mental Activities: Implication of Movement Involving Skeletal Muscle," *American Journal of Physiology* 91 (1929): 597–608.

113 **Kai Miller and his colleagues:** Miller et al., "Cortical Activity During Motor Execution, Motor Imagery, and Imagery-Based Online Feedback."

114 **In the last five decades:** Jeannerod, "Mental Imagery in the Motor Context"; K. L. Lichstein and E. Lipschitz, "Psychophysiological Effects of Noxious Imagery: Prevalence and Prediction," *Behavior Research and Therapy* 20 (1982): 339–45.

115 **Imagery practice has significantly:** J. Schneider, W. Smith, and S. Whitcher, "The Relationship of Mental Imagery to White Blood Cell (Neutrophil) Function in Normal Subjects," paper presented at the 36th Annual Scientific Meeting of the International Society for Clinical and Experimental Hypnosis, San Antonio, TX, 1984; V. W. Donaldson, "A Clinical Study of Visualization on Depressed White Blood Cell Count in Medical Patients," *Applied Psychophysiology and Biofeedback* 25 (2000): 117–28, https://doi.org/10.1023/A:1009518925859.

117 **Jodie Harlowe and her:** Jodie Harlowe, Stephanie Farrar, Lusia Stopa, and Hannah Turner, "The Impact of Self-Imagery on Aspects of the Self-Concept in Individuals with High Levels of Eating Disorder Cognitions," *Journal of Behavior Therapy and Experimental Psychiatry* 61, no. 1 (2008): 7–13.

117 **studies in clinical psychology:** Emily Holmes, "Mental Imagery in Emotion and Emotional Disorders," *Clinical Psychology Review* 30, no. 3 (2010): 349–62.

117 **and performance psychology:** D. Feltz and D. M. Landers, "A Revised Meta-Analysis of the Mental Practice Literature on Motor Skill Learning," in *Enhancing Human Performance: Issues, Theories, and Techniques.*, eds. D. Druckmann and J. A. Swets (Washington, DC: National Academy Press, 1988), 61–101; Adam J. Toth, Eoghan McNeill, Kevin Hayes, et al., "Does Mental Practice Still Enhance Performance? A 24 Year Follow-Up and Meta-Analytic Replication and Extension," *Psychology of Sport & Exercise* 48 (2020): 101672, https://doi.org/10.1016/j.psychsport.2020.101672.

118 **"The emotional environment":** Bernie Siegel, *Peace, Love and Healing* (New York, Harper and Row, 1989), 35.

119 **"I knew I was":** Cal Botterill and Terry Orlick, *Visualization: What You See Is What You Get*, video, Coaching Association of Canada, 1988.

128 **envisioning from the internal:** Maamer Slimani et al., "Effects of Mental Imagery on Muscular Strength in Healthy and Patient Participants: A Systematic Review," *Journal of Sports Science and Medicine* 15, no. 3 (2016): 434–50.

133 **"Bianca Andreescu began":** Melissa Couto, "Canada's Bianca Andreescu Says Meditation, Visualization Formed Winning Mindset," Canadian Press, September 8, 2019.

149 **"I operate from a place":** Brian Hiatt, "Lady Gaga: New York Doll," *Rolling Stone*, June 11, 2009, https://www.rollingstone.com/music/music-news/lady-gaga-new-york-doll-244453/.

150 **"Jason knows that":** *This American Life*, episode 513, "129 Cars," produced and hosted by Ira Glass, broadcast December 13, 2013, https://www.thisamericanlife.org/513/transcript.

Chapter Five: Protecting Your Confidence Every Day, No Matter What

157 **"fed by previous successes":** Bruce Lee, *The Tao of Jeet Kune Do* (Burbank, CA: Ohara Publications, 1975), 68.

164 **"an individual with a":** Martin Seligman and Peter Schulman, "Explanatory Style Predicts Grades and Retention Among West Point Cadets" (unpublished paper, December 1990).

172 **A comprehensive meta-analysis:** Antonis Hatzigeorgiadis et al., "Self-Talk and Sports Performance: A Meta-Analysis," *Perspectives on Psychological Science* 6, no. 4 (2011): 348–56.

173 **"the existent evidence base":** David Tod, James Hardy, and Emily Oliver, "Effects of Self-Talk: A Systematic Review," *Journal of Sport and Exercise Psychology* 33 (2011): 666–87.

174 **In a 2014 study:** Anthony Blanchfield et al., "Talking Yourself Out of Exhaustion: The Effects of Self-Talk on Endurance Performance," *Medicine and Science in Sport and Exercise* 46, no. 5 (2014).

175 **"If I could go back":** Alex Hutchinson, *Endure: Mind, Body, and the Curiously Elastic Limits of Human Performance* (New York: William Morrow, 2018), 260.

178 **"When doubt seeps in":** Chael Sonnen and Uriah Hall, "Chael Sonnen Coaching in *The Ultimate Fighter* about The Doubt," uploaded to YouTube by AGEMO, February 16, 2013, originally broadcast on *The Ultimate Fighter*, season 17, aired February 2013 on FX, https://www.youtube.com/watch?v=k5M8CKDYwM4.

Chapter Six: Deciding to Be Different

209 **"recalling autobiographical memories":** Megan E. Speer and Mauricio R. Delgado, "Reminiscing About Positive Memories Buffers Acute

Stress Responses," *Nature Human Behaviour* 1, no. 5 (2017), doi:10.1038/s41562-017-0093.

209 Other studies indicate: Adrian Dahl Askelund, Susanne Schweizer, Ian M. Goodyer, and Anne-Laura van Harmelen, "Positive Memory Specificity Is Associated with Reduced Vulnerability to Depression," *Nature Human Behaviour* 3 (2019): 265–73.

209 joy and excitement: D. P. Templin and R. A. Vernacchia, "The Effect of Highlight Music Videotapes on Game Performance in Intercollegiate Basketball Players," *The Sport Psychologist* 9 (1995): 41–50; D. Gould, R. C. Eklund, and S. S. Jackson, "1988 U.S. Olympic Wrestling Excellence: Mental Preparation, Precompetitive Cognition, and Affect," *The Sport Psychologist* 6 (1992): 383–402.

211 "You have to step": "Helen Maroulis: The Next Generation," History NOW, episode 5, April 1, 2016, https://www.youtube.com/watch?v=yVoUagBUxXA.

213 "The thing I've always": "Tiger Woods interview with Oprah Winfrey After 1997 Masters Victory," uploaded to YouTube by GOATvzn, 2018, originally broadcast on *The Oprah Winfrey Show*, April 24, 1997, https://www.youtube.com/watch?v=z36FCcr9j2w.

220 "That's exactly what you": Eli Manning on the *Michael Kay Show*, ESPN Radio, February 7, 2012.

Chapter Seven: Entering the Arena with Confidence

227 "There's no substitute": Dan McGinn, *Psyched Up: How the Science of Mental Preparation Can Help You Succeed* (New York: Penguin/Portfolio, 2017), 236.

228 "I come down to": Gary Cohen, "Billy Mills—September 2014," GaryCohenRunning.com, http://www.garycohenrunning.com/Interviews/Mills.aspx.

229 "This is where I begin": Andre Agassi, *Open: An Autobiography* (New York: Knopf Doubleday, 2009), 9.

234 "This guy's mind is": E. M. Swift, "Paul Kariya," *Sports Illustrated*, February 22, 1993, https://vault.si.com/vault/1993/02/22/paul-kariya.

239 "his entire energy field": Susan Casey, "Gold Mind," *Sports Illustrated*, August 18, 2008.

241 To transition to his: Mark McLaughlin and Shawn Coyne, *Cognitive Dominance: A Neurosurgeon's Quest to Outthink Fear* (New York: Black Irish LLC, 2019).

245 "I get up": Steven Pressfield, *The War of Art* (New York: Black Irish LLC, 2002), vii.

250 "That was the most": "Helen Maroulis: 'I Am Enough,'" video, NBC Olympic Coverage, September 8, 2016, http://archivepyc.nbcolympics.com/video/helen-maroulis-i-am-enough.

Chapter Eight: Playing a Confident Game from Start to Finish

256 "Overwhelming support in the": B. Hatfield and S. Kerrick, "The Psychology of Superior Sport Performance: A Cognitive and Affective Neuroscience Perspective," in *Handbook of Sport Psychology Research* (3rd ed.), eds. Gershon Tenenbaum and Robert C. Eklund (New York: John Wiley and Sons, 2007).

257 "From a neuroscience point": Brad Hatfield on FitTV episode, 2002.

262 Sport psychology research: S. G. Ziegler, "Effects of Stimulus Cueing on the Acquisition of Groundstrokes by Beginning Tennis Players," *Journal of Applied Behaviour Analysis* 20 (1987): 405–11; B. S. Rushall et al., "Effects of Three Types of Thought Content Instructions on Skiing Performance," *The Sport Psychologist* 2 (1988): 283–97; D. Landin and E. P. Hebert, "The Influence of Self-Talk on the Performance of Skilled Female Tennis Players," *Journal of Applied Sport Psychology* 11 (1999): 263–82; C. J. Mallett and S. J. Hanrahan, "Race Modelling: An Effective Cognitive Strategy for the 100 m Sprinter?," *The Sport Psychologist* 11 (1997): 72–85; N. Zinsser, L. Bunker, and J. M. Williams, "Cognitive Techniques for Building Confidence and Enhancing Performance," in *Applied Sport Psychology: Personal Growth to Peak Performance* (5th ed.), ed. J. M. Williams (Boston: McGraw-Hill, 2006), 349–81.

263 "You never want to": Bob Bowman interview, "The Mental Game," uploaded to YouTube by motiv8ireland, May 10, 2008, https://www.youtube.com/watch?v=lFbOKJEIJ00.

266 "Focusing on my breathing": B. Vranich and B. Sabin, *Breathing for Warriors* (New York: St. Martin's Publishing, 2020), 217.

269 "It's almost as if": *Tiger: The Authorized DVD Collection* (Burbank, CA: Buena Vista Home Entertainment, 2004) DVD.

279 "I actually have to": "Olympic Wrestler Helen Maroulis Fights Like a Girl," directed by Liza Mandelup, *Vogue*, August 10, 2016, https://www.vogue.com/video/watch/wrestler-helen-maroulis-rio-2016-summer-olympics-how-to-fight.

Index